# METALLURGY

## A BRIEF OUTLINE

OF THE

## MODERN PROCESSES FOR EXTRACTING THE MORE IMPORTANT METALS

BY

### W. BORCHERS

KGL TECH HOCHSCHULE, AACHEN

### AUTHORIZED TRANSLATION FROM THE GERMAN

BY

### WILLIAM T. HALL AND CARLE R. HAYWARD

*Instructor in Chemistry*        *Instructor in Metallurgy*

MASSACHUSETTS INSTITUTE OF TECHNOLOGY

*FIRST EDITION*

FIRST THOUSAND

NEW YORK

JOHN WILEY & SONS

LONDON: CHAPMAN & HALL, LIMITED

1911

THE SCIENTIFIC PRESS
ROBERT DRUMMOND AND COMPANY
BROOKLYN, N. Y.

# PREFACE

THE purpose of the book is to present, in concise form, the different processes used for extracting the important metals from their ores, and for refining them. Both the student and the practical engineer can thus get a broad view of modern metallurgical operations, without taking the time required for a study of the more detailed treatises on the subject. Whenever possible, self-explaining illustrations of apparatus have been used in place of descriptions. The microphotographs are from articles by P. Goerens, Aachen, and W. Campbell, New York, published in Metallurgie.

The present translation is commended to the public in the hope that such a summary of metallurgical processes will prove of assistance to English-speaking students in pointing the way to further study. The translators are indebted especially to Professor H. O. Hofman, for his kind advice and assistance in reading the proof-sheets.

WILLIAM T. HALL
CARLE R. HAYWARD

MASSACHUSETTS INSTITUTE OF TECHNOLOGY,
Boston, Mass., February, 1911

# TABLE OF CONTENTS

# METALLURGY

## GOLD

### Sources

**Natural Sources:**

NATIVE; alloyed with Pt, Ag, Hg, or Fe, as

VEIN GOLD, in quartz and other ancient rocks (primary deposits), and

PLACER GOLD, alluvial gold or gold dust (secondary deposits).

MINERALIZED, rarely; then associated chiefly with

TELLURIUM; also with many sulphides, particularly copper ores, whether as native gold or as sulphide, is difficult to determine.

**Other Sources:**

MATTES containing precious metals, especially from copper smelting and from pyritic smelting.

ALLOYS (crude metals) from other smelting operations.

SCRAP from the manufacture of jewelry, and

OLD GOLD and plated ware.

## (A) Extraction

**Ore Dressing:**

**Purely Mechanical Dressing,** usually the first method employed for developing a newly-discovered deposit.

**Electro-magnetic Concentration,** used as a preliminary treatment of sands carrying magnetite.

**Solution of Gold in Metals:** the metals used as solvents are copper, lead, mercury, and silver.

**In Copper.** The formation of a gold-copper alloy results in most copper smelters when gold-bearing copper ores, or smelter products, are smelted for copper; in pyritic smelting when gold-bearing quartz is used as lining for blast furnace, converter, or hearth of the reverberatory furnace; in the copper crust that is removed when liquating lead that carries precious metals; and, finally, in working up old gold and gold-plated ware, or residues from the manufacture of jewelry.

**In Lead.** A gold-lead alloy is formed by melting gold-bearing alloys together with lead ores, or metallic lead, and is obtained in the subsequent separation processes (see Silver).

**In Mercury (Amalgamation).** The metal as it exists in the free state, or as it is set free from its chemical compounds, is dissolved in mercury, the resulting amalgam is cleaned by mechanical treatment, and the precious metal is recovered by distilling off the mercury. Since silver compounds usually take part in the reactions, an outline of the process will be given under Silver.

**In Silver.** A gold-silver alloy is formed by most of the methods used for treating ores containing silver and gold; also by direct fusion preparatory to the parting of Au and Ag.

**Chemical Solution and Precipitation of Gold:**

**Chlorination.** The free gold, as it occurs native or as it is formed by a preliminary roasting, is converted into chloride by the action of chlorine gas upon the moist ore (Plattner) or by the action of dissolved chlorine (acid and bleaching powder) upon the finely-divided ore. The latter process, as perfected by Thies, Rothwell, and others, is the one most used in modern plants.

SUITABLE ORES: those which contain little or none of substances that absorb chlorine (Cu, Pb, Zn, As, Sb, Te, S, CaO, MgO). Some ores require a preliminary roast. The different steps are as follows:

1. CRUSHING, accompanied by sampling. The ore is crushed

to a maximum size of 60 mm. and screened to separate the fines (under 15 mm.) The over-size is put through

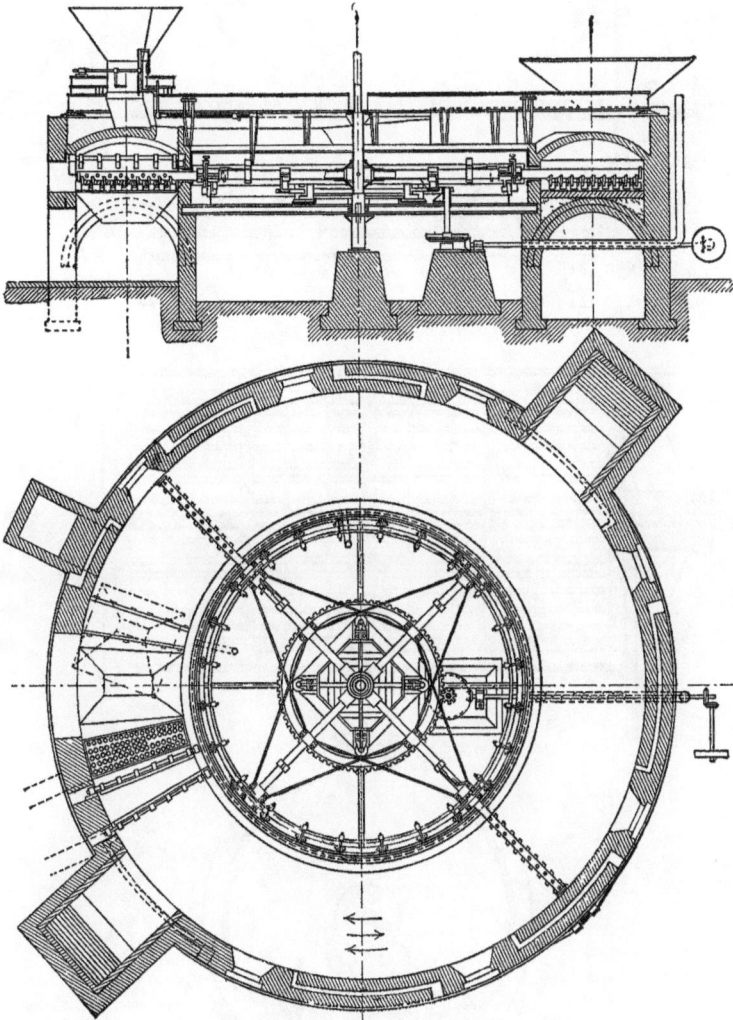

FIGS. 1 and 2.—Pearce Turret Furnace ($^1/_{150}$).

rolls to reduce it. A sample is taken at this point, and the furnace charge is prepared.

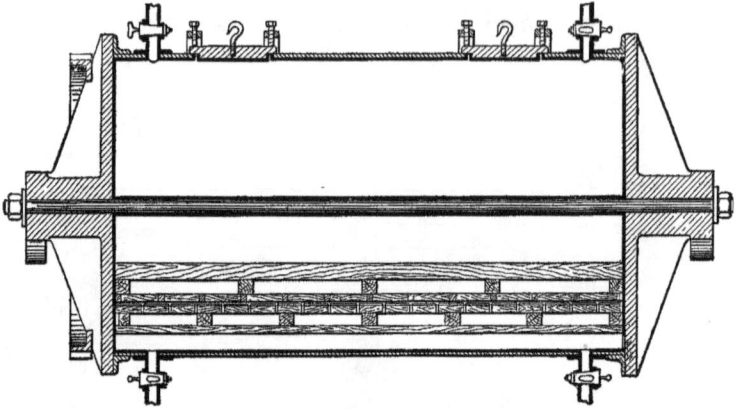

FIG. 3.

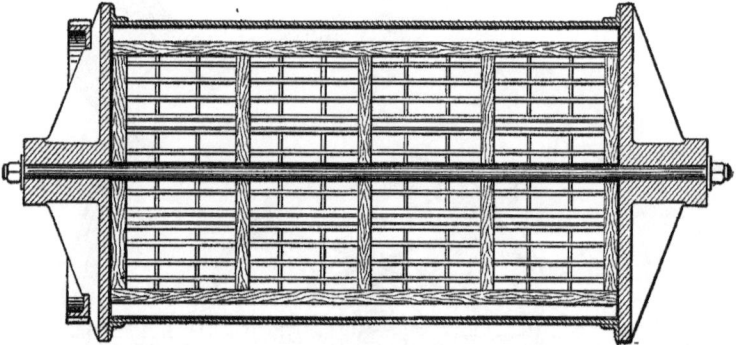

FIG. 4.

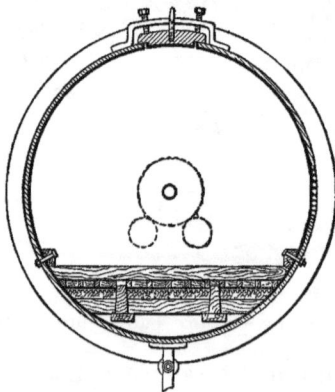

FIG. 5.—American Chlorination Barrel (Scale 1 : 60).

2. DRYING in a reverberatory with mechanical stirrers or with a rotating hearth.

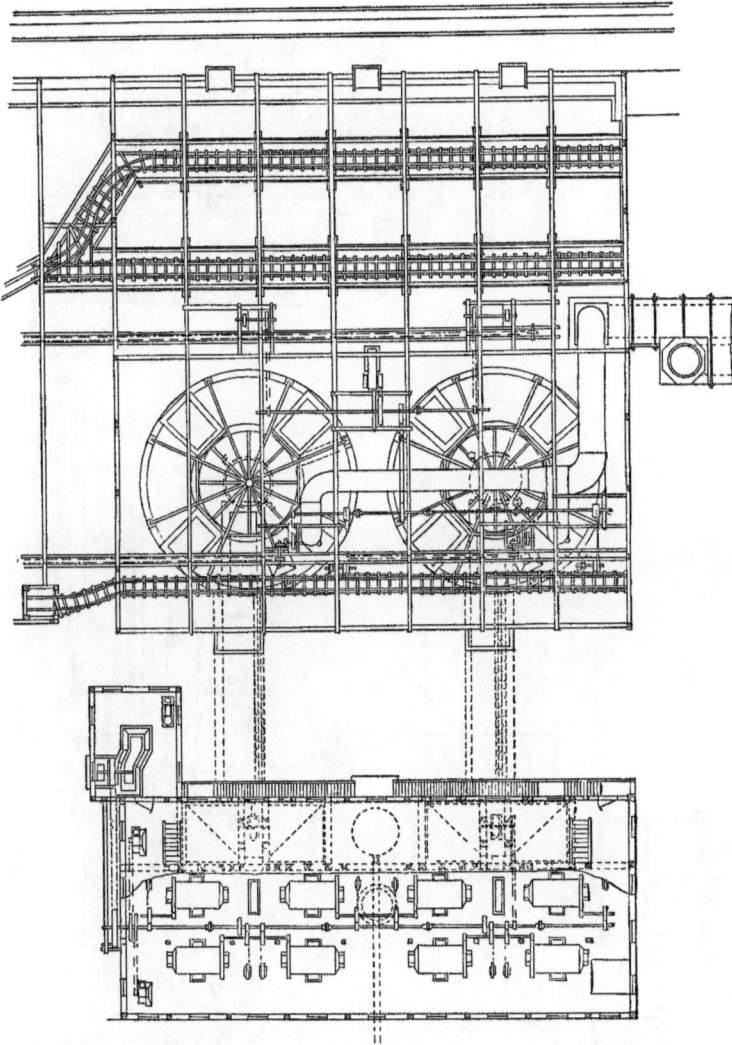

FIG. 6.—Plan of a Chorination Plant.

3. FURTHER CRUSHING with fine rolls (7 to 20 meshes per inch).

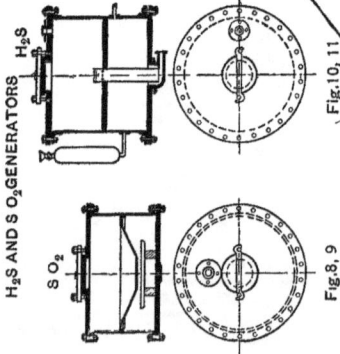

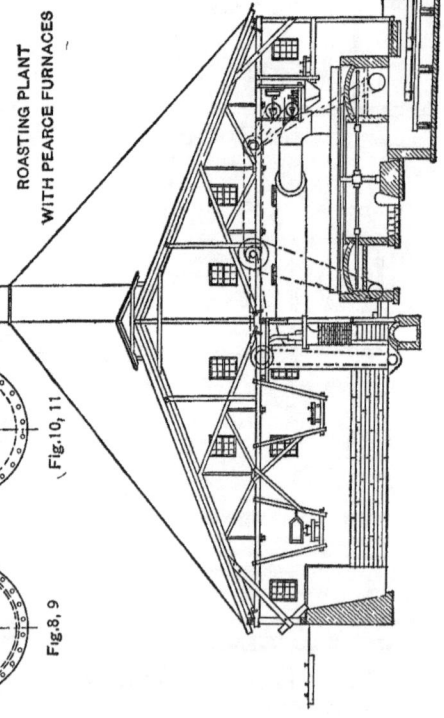

CHLORINATION AND
PRECIPITATION PLANT

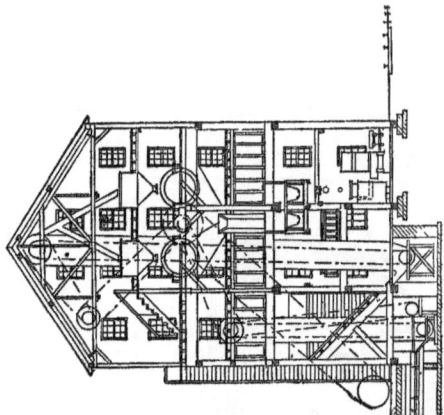

Fig.7

ROASTING PLANT
WITH PEARCE FURNACES

H₂S AND S O₂GENERATORS

H₂S

Fig.10, 11

SO₂

Fig.8, 9

4. ROASTING in Wethey, Pearce, or similar furnace. Capacity of the Pearce furnace: 100 tons* of ore require 1070 to 1500 sq.ft. of hearth area, 75 sq.ft. of grate area and 10 to 15 tons of coal. Temperature 400° to 800° C. (750° to 1550° F.).

5. CHLORINATION takes place in revolving, lead-lined ($\frac{3}{8}$ in. thick) barrels made of sheet iron ($\frac{1}{8}$ in. thick) and provided with a filter. Diameter 5 ft. 5 in. to 5 ft. 7 in.; length 14 to 16 ft.; capacity 10 tons ore and 5 tons water. (Figs. 3, 4, and 5.) For the generation of chlorine, 200 to 400 lbs. of $H_2SO_4$ and 100 to 200 lbs. of bleaching powder are required. Time: 2.5 to 4 hours with 3 to 5 revolutions per minute. A smaller barrel, about 4 ft. 7 in. long and 3 ft. 3 in. in diameter, has a capacity of 1 ton of ore and 160 to 200 gals. of water. For generating chlorine 18 to 40 lbs. bleaching powder and 20 to 50 lbs. $H_2SO_4$ are added. Twelve revolutions per minute.

6. FILTRATION inside the barrel, and washing with water under 2 to 3 atmospheres pressure (2.5 to 4 hours).

7. PRECIPITATION. A preparatory treatment with $SO_2$ to react with free chlorine is often used to-day when necessary. In lead-lined vats the gold is precipitated with $H_2S$. (Chlorine plant, $SO_2$ and $H_2S$ generators—see Figs. 6–11).

8. FILTRATION: in filter presses. The filtrate is refiltered through sand filters.

9. THE FILTER CAKES, together with the cloths, are placed in iron dishes, roasted in a muffle furnace and mixed with soda, borax and niter (Figs. 12–15).

10. FUSION in graphite crucibles.

11. POURING into molds.

**Cyanide Leaching and Precipitation.** This originated with the discovery by Elsner in 1844 of the solubility of gold in alkali cyanide solutions, in the presence of either free

---

* The tons referred to throughout the book are *metric tons* (2200 lbs.).

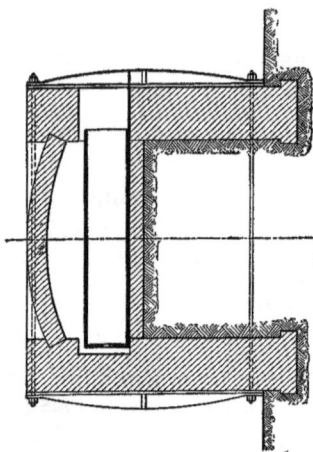

Fig. 14.

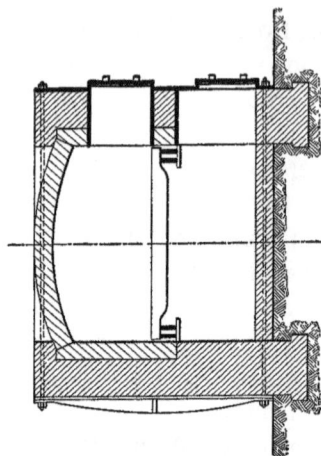

Fig. 15.

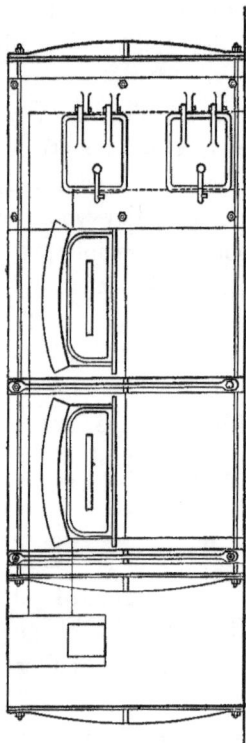

Fig. 12.

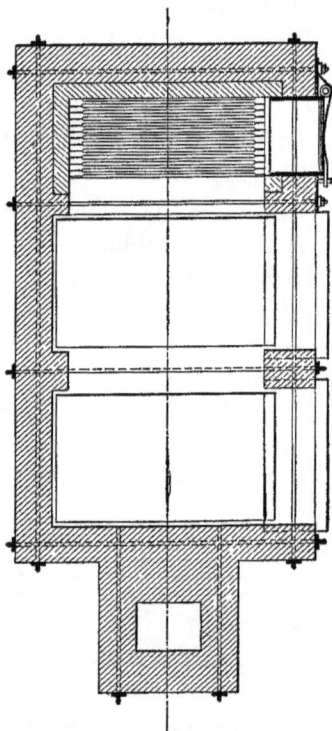

Fig. 13.

oxygen or oxidizing agents. According to Elsner the equation is,

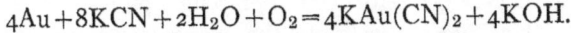

$$4Au + 8KCN + 2H_2O + O_2 = 4KAu(CN)_2 + 4KOH.$$

According to Bodländer (1896) and Christy (1900), the gold-potassium cyanide is formed by two reactions, viz.:

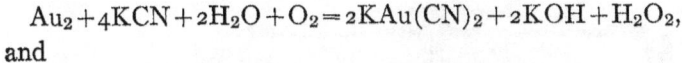

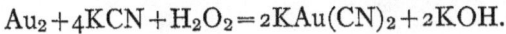

$$Au_2 + 4KCN + 2H_2O + O_2 = 2KAu(CN)_2 + 2KOH + H_2O_2,$$
and
$$Au_2 + 4KCN + H_2O_2 = 2KAu(CN)_2 + 2KOH.$$

When the ore is at all suitable, the cyanide leaching process is the ideal supplement to amalgamation. Amalgamation requires a large grain that will sink quickly to the bottom and be taken up by the mercury as the slime passes through the apparatus. The cyanide leaching, on the other hand, requires finely divided gold for a quick completion of the solution process because, according to the law of mass-action, the speed of any reaction depends, up to the velocity constant that is characteristic of each reaction, on the number of impacts of the molecules taking part in the reaction, and therefore upon the number of particles moving in the free state.

SUITABLE ORES: ores which are free from substances that destroy cyanide or precipitate gold (iron salts, organic acids, and metals having a high solution tension such as Cu, Zn). Ores containing Te, As and Sb require roasting. Acid ores require the addition of a neutralizing substance (milk of lime). Ores containing reducing agents require the addition of oxidizing agents (potassium ferricyanide, potassium permanganate, peroxides, bromine, etc.).

The old method, which is partly used to-day, is as follows, including the preparation of the ore:

1. BREAKING the ore in rock breakers.
2. FURTHER CRUSHING and beginning of amalgamation in stamp mills.

3. AMALGAMATION in shallow troughs lined with amalgamated copper plates.

4. CONCENTRATION on bumping tables. The object of this is to separate out the concentrates (heavy pyrite and

FIG. 16.

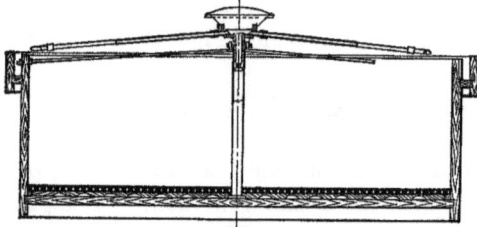

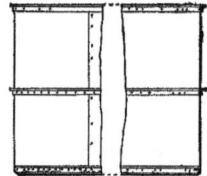

FIG. 18.—Details of Iron Work.

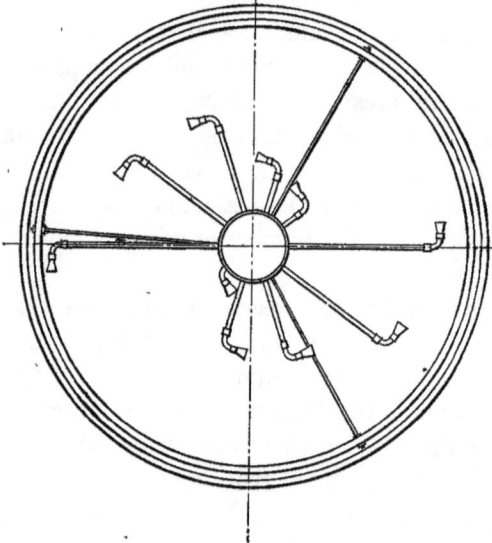

FIG. 19.—Details of Iron Work.

FIG. 17.—Vat with Filter Bottom and Butters' Distributor.

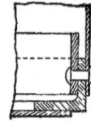

large grains of sand). These, after roasting if necessary, are treated with cyanide solution in the leaching vats.

5. SEPARATION OF THE TAILINGS into sands and slimes by means of *spitzlutten* or settling tanks (Fig. 16–19). The

spigot products from the *spitzlutten* and the settlings from
the settling tanks are treated in percolation vats and the
slimes in agitation vats.  If the ore is acid, milk of lime
or alkali solution is added.

6. DRAINING the sands or concentrates in the percolation
vats.

7. TREATMENT WITH STRONG CYANIDE SOLUTION (see
table below).

8. DRAWING OFF of the cyanide solution and aërating for
several hours.

9. WASHING WITH WEAK SOLUTION: this is repeated several times if necessary, the ore being allowed to aërate for
a short time between treatments.

10. WASHING WITH WATER, twice if necessary.

11. THE SLIMES after the addition of the cyanide˙ are put
into vats (usually by means of centrifugal pumps) and
agitated either with stirrers, or by blowing in air, in order
to extract the gold.

12. SEPARATION OF THE SLIME AND SOLUTION by decantation and filter-pressing.

A summary of the separate treatment of concentrates, sands
and slimes is shown in the following table:

| Materials. | Strength of Solution | Time of Leaching |
|---|---|---|
| Concentrates..................... | 0.01–0.1% KCN | 2–3 hours |
| Sands........................ | 0.01% KCN | 5–7 days |
| Slimes....................... | 0.01% KCN | 8–12 hours. |

**Recent Leaching Methods.**  The higher extraction that is
possible from slimes and the rapidity with which the gold
is dissolved has led, first in the Kalgoorlie district, Australia,
and subsequently in South Africa and in the United States,
to sliming the tailings, as they come from the stamp mill and
amalgamated plates, and leaching them in this form.  The
method was carried out successfully in the wet tube mill first

A Modern American Cyanide Plant.

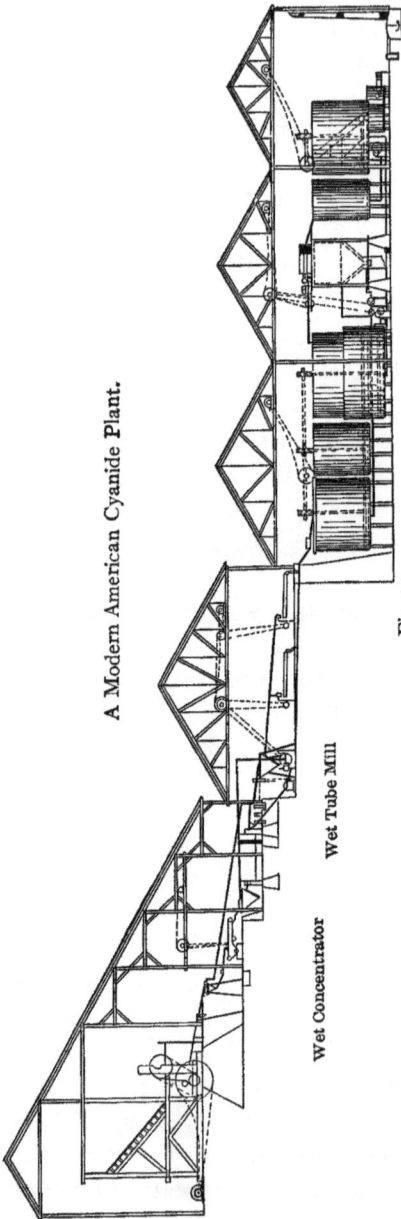

Wet Concentrator    Wet Tube Mill

Fig. 20

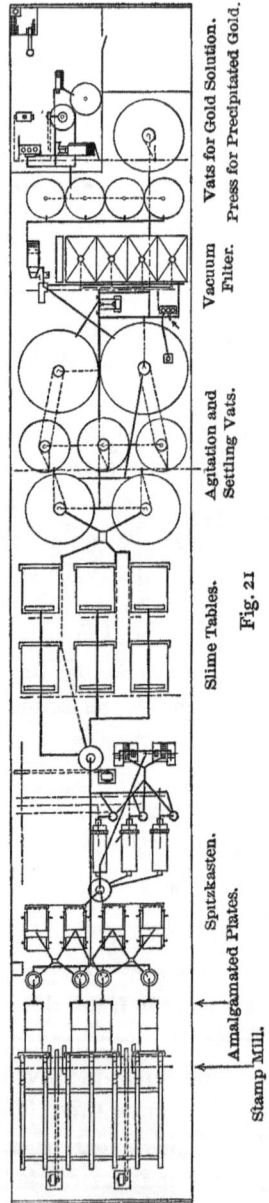

Stamp Mill.    Amalgamated Plates.    Spitzkasten.    Slime Tables.    Agitation and Settling Vats.    Vacuum Filter.    Vats for Gold Solution. Press for Precipitated Gold.

Fig. 21

built by the F. Krupp Co., using the following process as developed by Diehl:

1. PRELIMINARY CRUSHING in rock breakers.
2. FURTHER CRUSHING and beginning of amalgamation in the stamp mills.
3. PLATE AMALGAMATION.
4. FURTHER PULVERIZING in wet tube mills. These are rotating iron drums lined with steel plates and filled with

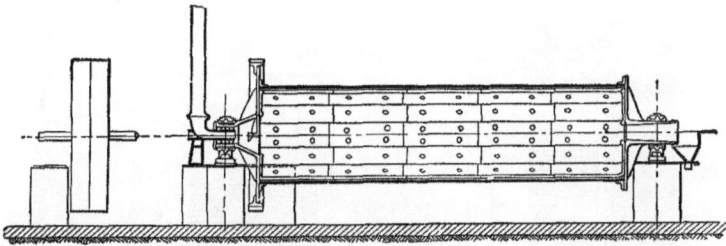

Fig.22

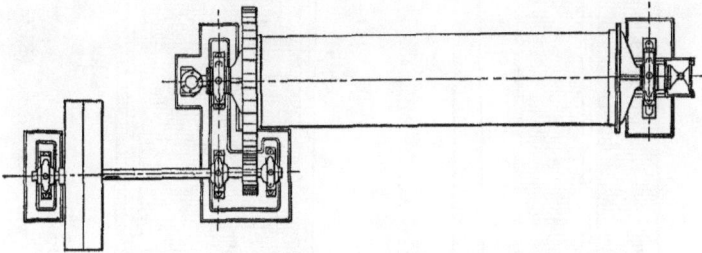

FIG. 23.—Krupp Wet Tube Mill. Scale, 1:100.

rounded flint stones. The inside dimensions of the drums are: diameter, 3.5 to 5.5 ft.; length, 13 to 20 ft. Revolutions, 30-24 per minute. Power required, 24-28 H.P. Capacity: One 24-H.P. mill will grind 65 tons of ore in 24 hours to 100 mesh size.

5. SEPARATION OF THE SANDS in the *spitzlutten* and carrying them back into the tube mill.
6. AGITATION WITH CYANIDE, adding some oxidizing agent if necessary. By using 0.2% KCN with 0.05% Br.,

FIG. 26.—Front Elevation.

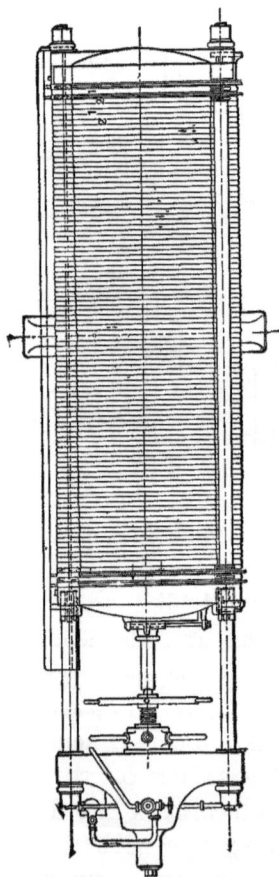

FIG. 27.—End Leaf.

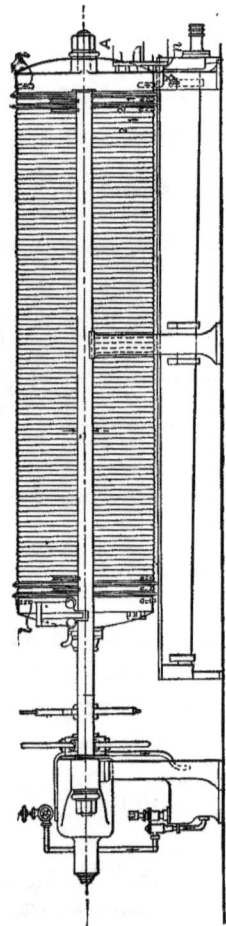

FIG. 24.—Side Elevation.

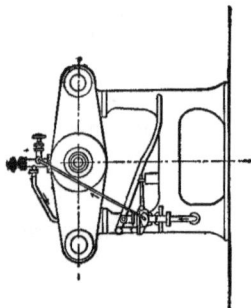

FIG. 25.—Plan.

Filter Press with Iron Frames (Klein, Schanzlin & Becker, Frankenthal).

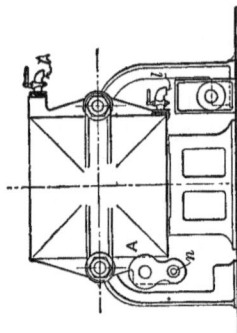

Diehl has even leached telluride ores successfully without roasting.

7. RECOVERY OF THE GOLD SOLUTION by decantation and filtration, or by filtration alone; in filter presses (Figs.

Fig. 28 FILTER FRAME

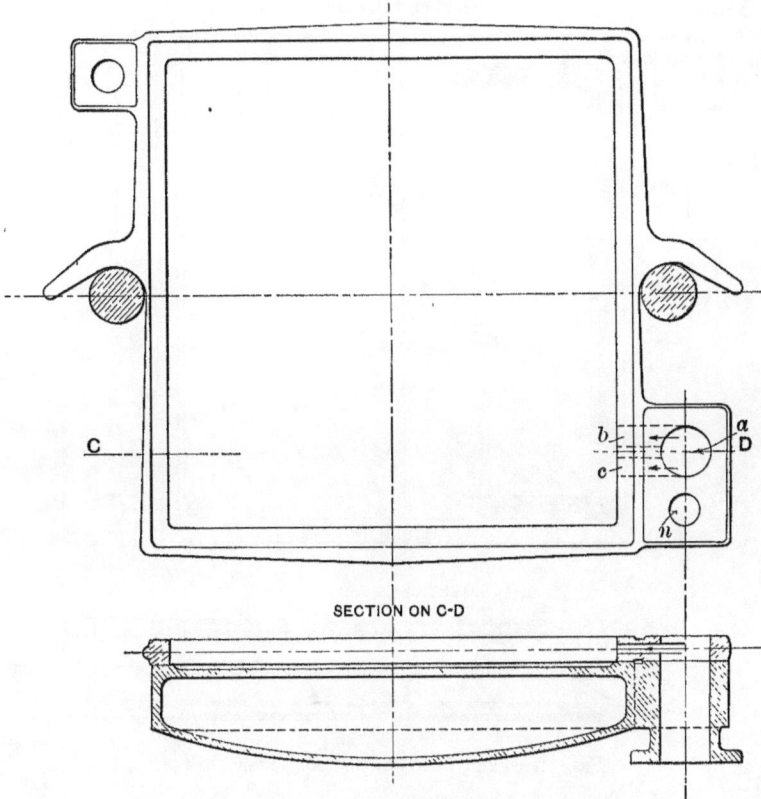

SECTION ON C-D

FIG. 29.—Head Filter Leaf with First Filter Frame. Scale, 1:15.

24–34) Butters' filters (Figs. 35–36) or Moore's filters (Metallurgie, 4, 559).

The precipitation of gold from the cyanide solutions is effected either by clean zinc shavings (old method), by lead coated zinc shavings (method of Betty and Carter), by electricity (method of Siemens and Halske), or by electricity followed by zinc treatment.

PRECIPITATION BY ZINC is accomplished by passing the gold solution through long wooden boxes, in which the zinc rests on filter bottoms in small cells with double partition

Fig. 30 FILTER LEAF

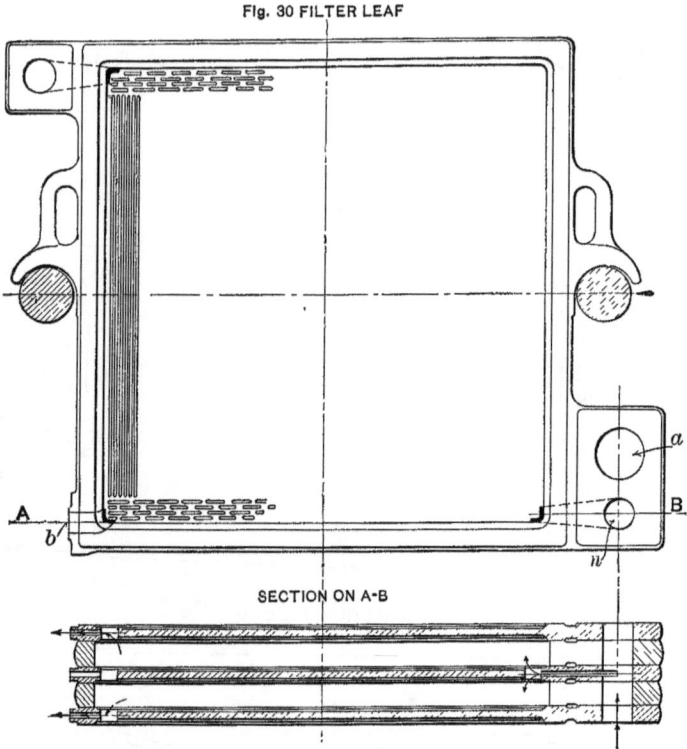

SECTION ON A-B

FIG. 31.—Filter Leaf and Frame.   Scale, 1 : 15.

walls. The solution rises through one cell, descends between the partitions, rises through the next cell, and so on through the vat. When the cells are cleaned up, the larger pieces of zinc are removed mechanically and the remainder dissolved by treatment with sulphuric acid. The precipitate is then washed, dried, melted with lead, and cupelled (see Cupellation, under Silver).

FIG. 32.

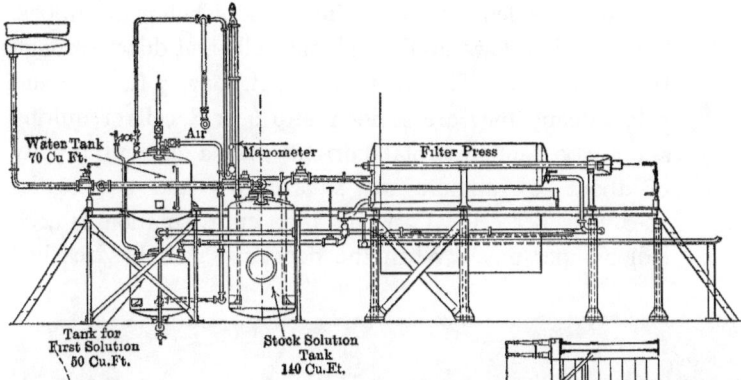

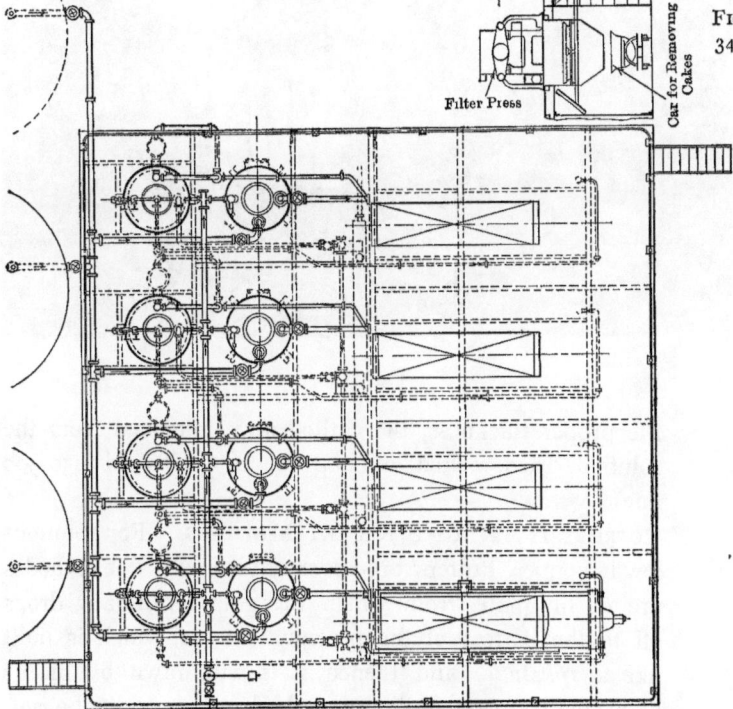

FIG. 33.—Filter Pressing Plant of the London and Hamburg Gold Recovery Co., Ltd., London, Eng.

ELECTROLYSIS, OLD PROCESS OF SIEMENS AND HALSKE.
Anodes: iron plates. Cathodes: strips of lead, suspended
in a long wooden vat divided into cells by hollow partitions.
The solution flows up through the cells and down through
the partitions. The vats are 23 ft. long, 5 ft. wide and
3 feet deep; they are divided into 6 or 8 cells requiring
about 100 amperes total current with a current density
of about .05 ampere per square foot and a potential
of 2 volts per cell. Capacity 1800 cu.ft. solution (13,000
gallons) per day. When the deposited gold has reached

FIG. 35.

the proper thickness, the cathodes are removed from the
solution, dried, melted, and cupelled for crude gold (850-900
fine).

ELECTROLYSIS BY MORE RECENT METHODS. For solutions
low in copper, Butters' process may be used. The cathodes
are of tin plate, from which the gold continually drops
off to the bottom of the precipitation vat, which is built
like a *spitzlutte*, and thence it is withdrawn by means
of a small stream of liquid and filtered outside the vat.
Since with solutions running high in copper the deposit
is more dense, Charles P. Richmond (Metallurgie 4, 502)

has gone back to lead cathodes. These, when covered
with a gold or copper deposit of proper thickness, are

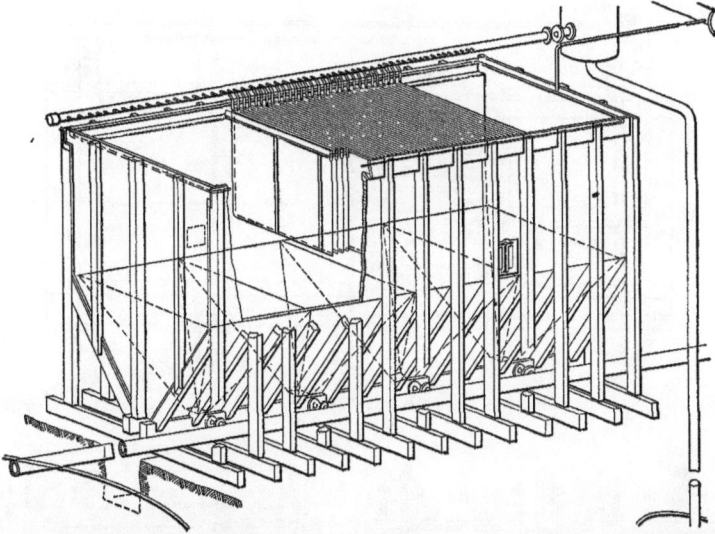

FIG. 36.—Butters' Vacuum Filter. Frames made of iron rods covered with
canvas. The spaces between are filled with cocoa matting. The perforated
iron pipes are connected to suction tubes.

Operation of the filter: The vat is filled with slime and suction applied.
The solution is drawn through the hollow frame until the cake on the outside of
the filter becomes so thick that filtration is difficult. The slime is then withdrawn
from the vat and water admitted for washing the cakes. After washing, the cakes
are removed from the filter by applying pressure instead of suction.

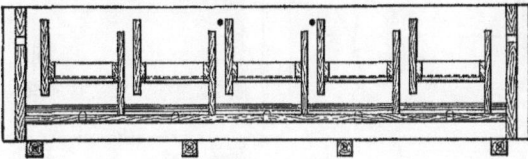

FIG. 37.                        FIG. 38.

Cross-sections of Zinc Boxes.

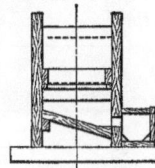

removed from the solution, enclosed in filter bags, and
placed in an acid bath, where they are used as anodes.
The gold remains behind as slime, while copper is dissolved

and goes to the cathode, but if a high current density is used it falls off and is recovered as a slime similar to cement

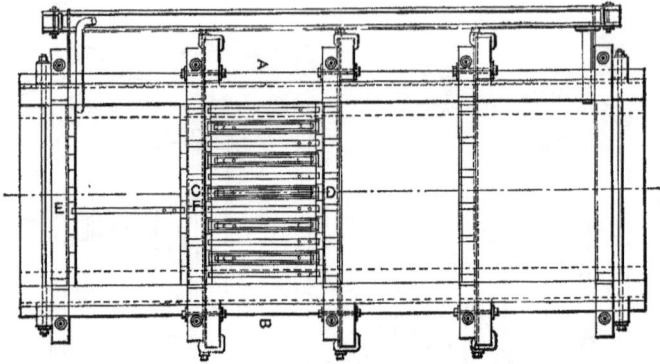

FIG. 39.—Plan.

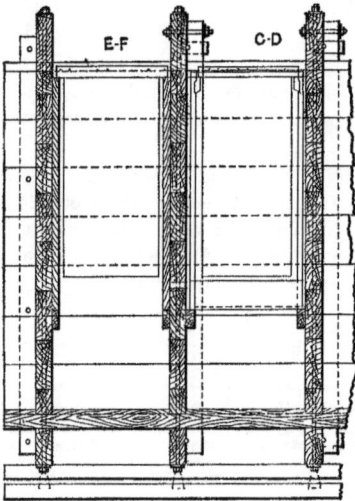

FIG. 40.—Sections *EF* and *CD*

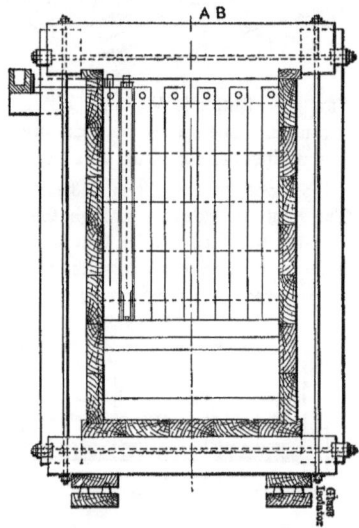

FIG. 41.—Section *AB*.

copper. The gold which remains as slime in the bags is dried and fused as described above.

**Parting:** gold as a residue. Most of the so-called *parting processes*

consist in dissolving away, either in aqueous solutions or by
fusion, the metals alloyed with the gold, leaving the latter

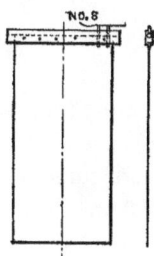

FIG. 42.—Anode.    FIG. 43.—Cathode: in the    FIG. 44.—Wooden
                   Second Part of the Process:         Frame.
                   Anode.

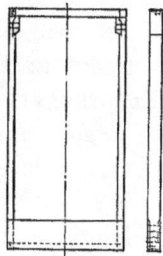

behind.   Of the many methods that have been devised, the
following are still used in practice:
　　Nitric acid parting, inquartation
　　Sulphuric acid parting,　　　　　　　}Described under Silver.
　　Fusion with sulphur.  Rössler process
　　Electrolytic parting and purification of other metals.   See
　　Silver, Copper and Nickel.

# (B) Gold Refining

**Solution and Precipitation of the Gold:**
　　**Chlorinating Solution.**  Chlorine and solutions evolving chlorine,
　　particularly aqua regia, are the principal agents used
　　in the final refining of gold.   For dissolving the gold, dilute
　　aqua regia suffices (2 parts concentrated HCl, 1 part concen-
　　trated $HNO_3$, 3 parts $H_2O$).   Apparatus: porcelain vats.
　　The gold solution is filtered to remove the insoluble residue
　　(AgCl) and the following operations are then carried out:
　　Precipitation of the gold by ferrous sulphate, filtration,
　　washing, drying, and melting in a graphite crucible under
　　a cover of glass and borax.   If platinum is present, it is

precipitated by adding metallic iron to the solution from which the gold has been removed.

**Electrolytic Parting,** used for refining gold that contains platinum.

ANODES: gold platinum alloy.

CATHODES: sheets of refined gold. Distance between electrodes 1.2 inches.

ELECTROLYTE: gold chloride solution containing 25–30 oz. of gold and 20–50 oz. HCl (sp.gr. 1.19) per cubic foot (7.5 gallons). If the anodes contain lead some $H_2SO_4$ is added. The large amount of free HCl in the electrolyte is required because the anode gold goes into solution only when the conditions are favorable for the formation of $HAuCl_4$, the reaction being: $Au + 3Cl + HCl = H(AuCl_4)$. Furthermore, the greater the current density, the larger the quantity of free acid that must be kept in the bath; a high current density is required for the rapid deposition of the gold. Electrolytic solution of the gold is increased by rise of temperature. The conditions for electrolysis are, therefore:

CURRENT DENSITY: 100 amperes per square foot, at times rising as high as 300 amperes per square foot.

POTENTIAL: 1 volt.

TEMPERATURE: 60° to 70°C. (140° to 158° F.).

ELECTROLYZING VESSELS: stone jars or porcelain beakers immersed in a water bath. An automatic method is used for replacing the water lost by evaporation from the water bath; the wash-water from the cathodes and anode mud is used chiefly for this purpose.

PLATINUM, although insoluble in HCl when used alone as an anode, goes into solution somewhat in the presence of gold, but it is not deposited on the cathode as long as the amount of platinum in the electrolyte does not become more than double the amount of gold. The platinum may be precipitated from a platinum-rich electrolyte by adding $NH_4Cl$, and the same is true of palladium. Silver is converted into insoluble AgCl.

THE PRODUCTS of electrolytic refining are gold, 999.8 to 1000 fine, platinum, and silver chloride.

**Dissolving or Slagging the Metals Alloyed with the Gold:**

SULPHURIC ACID REFINING consists in repeatedly boiling the gold with concentrated $H_2SO_4$. The further treatment of the gold residue is the same as described under Chlorinating Solution (p. 21).

NITRIC ACID REFINING consists in boiling repeatedly with concentrated nitric acid (see Silver, Ag-Cu parting).

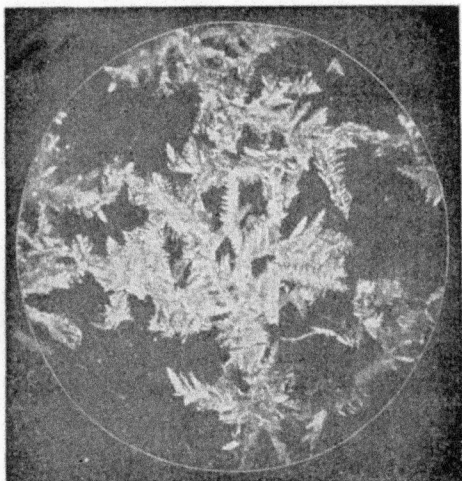

FIG. 45.—Crystals of Precipitated Gold (×35).

REFINING FUSION: gold containing small amounts of impurities, and also the precipitated gold as described on p. 21, which is to be melted and cast into bars, is fluxed with borax and glass, and if necessary with an oxidizing agent such as niter, and melted in graphite crucibles.

**Properties of Refined Gold:**

SPECIFIC GRAVITY: 19.3.

COLOR: yellow with brilliant lustre.

TENACITY: very tough, the most tenacious of metals.

FRACTURE: hackly.

STRUCTURE: (Fig. 45).

MELTING POINT: 1064° C. (1947° F.).

VAPORIZATION at 2000° C. (3632° F.).

ELECTRICAL CONDUCTIVITY: 0.6 to 0.7 referred to Ag = 1.

ALLOYS with most metals.  The alloys of Au with Pt, Ag, Hg, Cu, Pb, and Zn are important metallurgically.

CHEMICAL BEHAVIOR: not very active, has a low solution tension, and its compounds are readily broken down.  In extracting gold, the most important solvents are Cl, Br, and KCN.

# PLATINUM

## Sources

**Natural Sources:**

FREE, alloyed with the other platinum metals as well as with Fe, Au, and Cu.

MINERALIZED, occasionally as a compound with arsenic (Sperrylite, $PtAs_2$) but more frequently in Ni, Cu, and Au ores, although in small quantities.

**Other Sources:**

Smelter products: In matte and bullion from smelting platinum-bearing Ni, Cu, and Au ores.

## Extraction

**Concentration.** Wet, mechanical methods are nearly always utilized in the preliminary concentration of platinum-bearing gravel.

**Solution in Metals:**

**In Lead.** This is carried out occasionally by smelting platinum ores with PbO and PbS, whereby the Fe and Cu in the ores are converted into matte while the platinum metals enter the lead bullion, from which they are subsequently recovered by cupellation (see Silver Cupellation Methods, pp. 33–35).

In Nickel ⎫
In Copper ⎬ These processes take place along with the treatment of Ni, Cu, and Au ores. See these metals
In Gold ⎭ and their electrolytic separation.

**Chemical Solution and Precipitation of the Platinum:**

**Chlorination** in aqueous solution by means of mixtures evolving chlorine, especially aqua regia, is carried out in the refining of alloys. From the resulting solution the platinum is subsequently precipitated either by iron (see Pt-Au parting), or by ammonium chloride.

25

# Refining

**Chemical Solution and Precipitation** is utilized both in the direct working of the ores and in the parting of alloys or of crude platinum. The following treatments are given:

**Purification with HCl,** containing but a small amount of $HNO_3$, so that only the base metals are dissolved. This treatment is

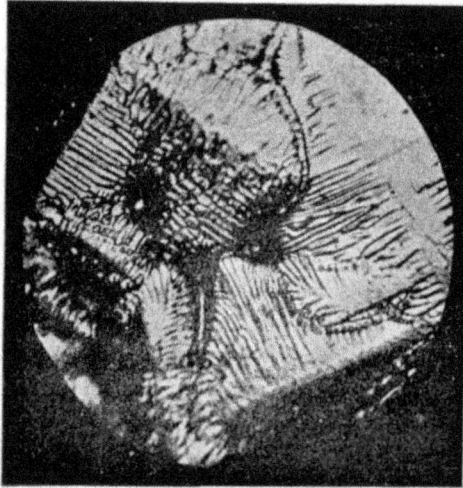

FIG. 46.—Surface of Fused Platinum ($\times 33$).

not given unless imperative. As a rule it is only necessary to carry out

**Chlorination** in aqueous solution as described above. If the solution carries gold, this can be precipitated by ferrous sulphate or by electrolysis. The separation of the accompanying platinum metals takes place usually by

**Partial Precipitation** with $NH_4Cl$, if necessary after the previous reduction of Pd and Ir chlorides from the " ic " to the " ous " state. Ammonium chloride precipitates the platinum as ammonium chloroplatinate, $(NH_4)_2PtCl_6$. The latter after

**Separation from the Solution** by decantation, filtering, washing, and drying is changed by

**Ignition** into platinum sponge:

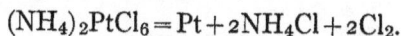

$$(NH_4)_2PtCl_6 = Pt + 2NH_4Cl + 2Cl_2.$$

**Melting.** In order to convert the precipitated powder or the sponge into a compact mass, it is heated by an oxyhydrogen blowpipe in a furnace made of limestone. This also removes the last traces of impurities.

**Properties of Platinum:**

SPECIFIC GRAVITY: 21.5.

COLOR: grayish white, brilliant lustre.

TENACITY: very ductile, very high tensile strength, capable of being drawn into the finest wires and the thinnest sheets (platinum foil).

STRUCTURE: see Fig. 46.

MELTING-POINT: 1745° C. (3173° F.).

ELECTRICAL CONDUCTIVITY: 0.08 (Ag = 1).

ALLOYS readily with most metals, especially when easily fusible, also with H.

CHEMICAL BEHAVIOR: not very active, low solution tension; compounds easily dissociated. For the extraction of platinum Cl is the most important solvent.

# SILVER

## Sources

**Natural Sources:**

NATIVE, alloyed with Au, Cu, Hg.

- MINERALIZED, principally in combination with the halogens and sulphur. Hornsilver, AgCl. Bromargyrite, AgBr. Iodargyrite, AgI. Silver glance, $Ag_2S$, occurs free, in sulpho salts (ruby silver, tetrahedrite), and as solid solution in sulphides.

**Other Sources:**

Burnt pyrite, matte from copper and lead smelting, slags, drosses, and alloys from metallurgical processes (black copper, base bullion, zinc skimmings, etc.).

OLD METAL and metallic scrap.

## (A) Extraction

**Concentration:** Nearly always carried on in connection with amalgamation (p. 35 et seq.).

**Solution of Silver in Metals:** Copper, lead, and mercury are commonly used as solvents.

IN COPPER: (see Gold and Copper).

IN LEAD: Silver, as it occurs free, or as it results from the decomposition of its compounds, is taken up by molten lead, and the two metals are subsequently separated. Materials poor in silver (under 10%) are smelted with lead ores (see Lead). Materials rich in silver (over 10%) are treated on a bath of molten lead (see Cupellation, pp. 33–35).

With lead as the carrier, most sulphide ores are satisfactorily treated, either directly or after roasting. Lead-free copper ores carrying precious metals are not used for this

purpose, but are worked by treating with them copper acting
as the carrier.   Ores, either raw or roasted, containing antimony and arsenic and various other materials not too rich in
copper are used.

**Operations:**

1. SMELTING of materials low in silver, with lead ore to

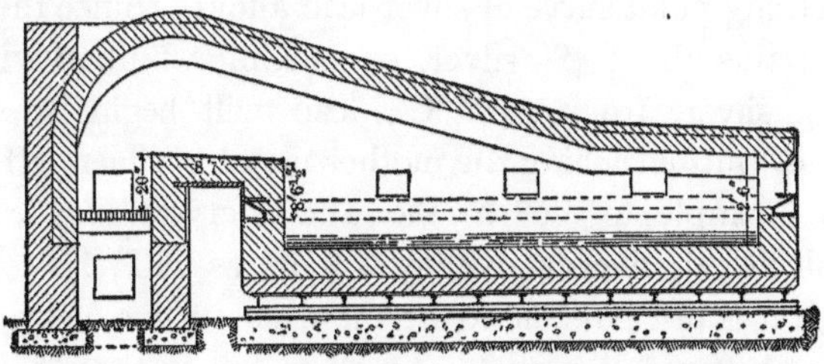

FIG. 47.—Refining Furnace.  Scale, 1 : 100.

obtain a base bullion low in silver (about 1% Ag).   Smelting directly to rich bullion leads to high slag losses.

2. REFINING THE BASE BULLION.  Melting in a deep-hearth
reverberatory furnace at a low temperature.   Removal of
the mechanically-enclosed impurities and of the difficultly-
fusible constituents; the
material    withdrawn,
usually copper-bearing,
is returned to the ore-
smelting furnace.

Heating in an oxidiz-
ing flame at a high tem-
perature   in   order   to
remove the larger part of
the antimony.  The oxi-
dation product (skim-
mings:   $Pb_3(SbO_4)_2 +$
$xPbO$), is refined under

FIG. 48.

a covering of charcoal in order to reduce part of the PbO.
The refined skimmings are smelted for hard or antimonial
lead, a Pb-Sb alloy with 14–20% Sb.

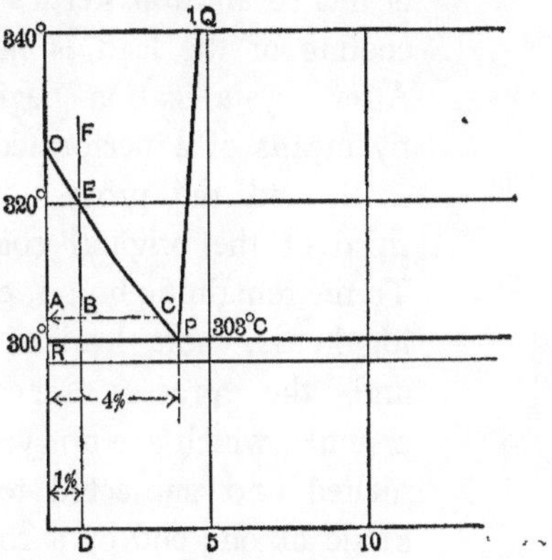

3. CONCENTRATION of the silver in a portion of the lead. This may be accomplished either by crystallization (Pattinson process) or by crystallization after adding zinc, and then separating the Ag as an Ag-Zn-Pb alloy (Parkes process). The enrichment of the bullion during the crystallization of the lead takes place according to the accompanying freezing-point curve of silver-lead alloys. Since the eutectic contains about 4% silver, on cooling a bullion with about 1% silver, from 340° C., lead will begin to separate at $E$, thus enriching the mother-metal in silver. If we cool, for example, to 306° C., the ratio of crystallized lead to the still liquid silver-lead alloy will be as $BC:AB$. In practice it is not possible to accomplish such a sharp separation between Pb and Pb-Ag alloy, for, in the skimming off of the crystallized lead or in the withdrawal of the liquid Pb-Ag alloy, some of the alloy adheres mechanically to the crystals, or small crystals remain suspended in the molten alloy. Two methods have been devised for the practical working of this process:

THE ORIGINAL PATTINSON PROCESS. The base bullion is melted in iron kettles and the dross skimmed. The cooling of the lead is hastened by spraying with water. After crystallization begins, the crystals are transferred by means of a perforated iron skimmer to a neighboring kettle and the process continued until only about one-third of the original contents remains in the first kettle. To the remaining liquid, either in the same or a neighboring kettle, more lead of the same silver value is added and the process repeated. Similarly the skimmed crystals, which are not yet sufficiently desilverized, are remelted and subjected to the same treatment. Thus the kettle at one end of a row will finally hold the enriched bullion (1.5 to 2% Ag) while the kettle at the other end will receive the desilverized lead.

THE LUCE-ROZAN PROCESS (Figs. 49–50, p. 32). The base bullion is melted in tilting kettles which discharge into the crystallizing kettle. Steam is blown through the latter until two-thirds of the lead has been crystallized and then the enriched bullion is removed by tapping. The crystallizing kettle is heated to remelt the lead crystals and refilled with bullion of the same silver value. The crystallization is repeated and the process continued until the lead remaining in the kettle is sufficiently desilverized. It is remelted and tapped into molds.

ZINC DESILVERIZATION is based upon the slight solubility of Zn in Pb, and conversely of Pb in Zn, and upon the high melting-point of the Zn-Ag-Pb alloy compared with that of the lead.

At 350° C., Pb dissolves 0.6 Zn and at 650° it dissolves 3% Zn.

The different operations of zinc desilverization are as follows:

The Cu is removed from the previously-purified, melted base bullion by adding to it a small amount of Zn (2%), allowing it to cool and removing the crystallized alloy. As soon as a sufficient quantity of copper skimmings have accumulated, they are liquated, thus producing (1) liquated bullion, which is returned to the bullion that has already been freed from Cu, and (2), a copper alloy low in silver (copper dust). The copper dust is melted and oxidized with steam, thus separating it into a rich bullion and oxides of Cu and other metals. These rich oxides are leached with $H_2SO_4$ so that a solution containing the sulphates of copper and of zinc results and a residue composed of particles of bullion which had been entangled in the oxides. The copper is precipitated from the solution by zinc scrap and the resulting solution of zinc sulphate yields zinc vitriol, $ZnSO_4.7H_2O$, on evaporation.

The bullion, freed from copper, is now heated up and the greater part of the silver removed by means of a larger addition of Zn (in the form of pure zinc, and skimmings poor in silver). The alloy which has crystallized

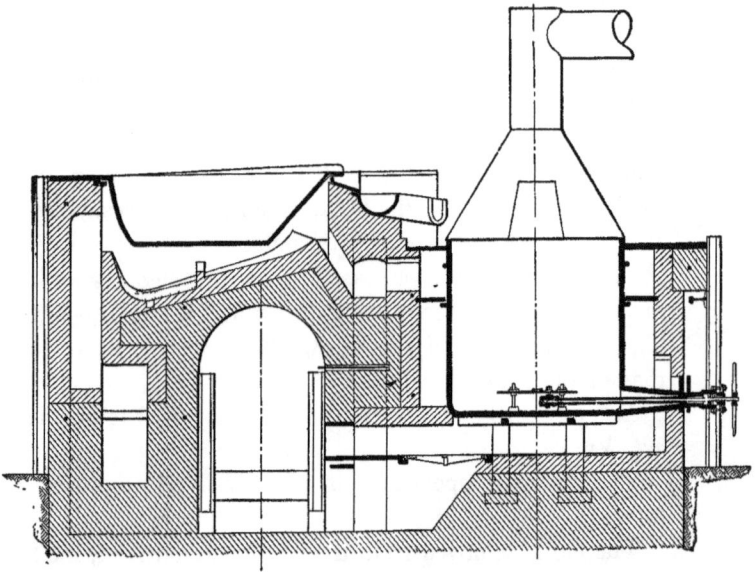

Fig. 49.

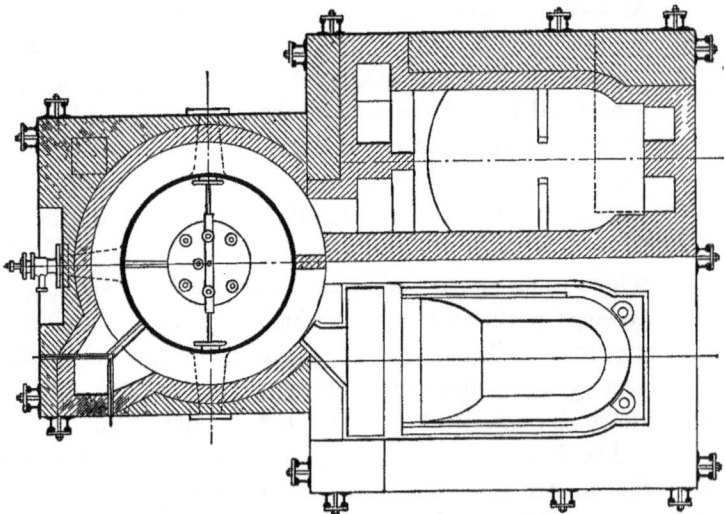

Fig. 50.—Pattinson Plant, using the Luce-Rozan Process.

out (zinc crust), is removed and liquated in another kettle. The liquated lead goes back to the desilverizing kettle. The residue (rich zinc crust) is distilled, giving rich bullion and zinc.

The desilverized lead is freed from zinc by blowing steam through it. The resulting mixture of zinc oxide and lead oxide which floats on the surface of the metal is skimmed, washed, and sold as pigment. If the lead still contains antimony, further oxidation is accomplished by means of an air current. The antimony is thus removed in the litharge which is formed at the same time. As a residue, there remains in the kettle pure lead, the so-called "soft lead."

As apparatus for desilverizing, hemispherical or flat cast-iron kettles with stirrers are used; for the distillation of the rich zinc crust, graphite or clay retorts; and for cupelling, reverberatories with hearths made of bone-ash or cement and lime. Lead-lined wooden vats serve for working up the so-called *rich oxides*, and for the poor oxides resulting from the dezincification of the lead, a jig is used together with a series of wooden slime tanks.

The rich bullion is converted, by cupelling, into litharge and crude silver, both the English and German methods being used. In the German process large reverberatory furnaces are used which take the entire charge at the beginning of the operation. In the English process, small reverberatory furnaces are used, the hearths being kept full during the operation by repeated additions of bullion until, finally, they are full of crude silver.

If liquated bullion is not used, the first litharge drawn off will be impure. After the removal of this, silver scrap may be added to the bullion. The base metals are oxidized and the precious metals dissolved in the bullion. Now begins the cupellation proper, namely the oxidizing smelting of the lead whereby the resulting litharge is constantly allowed to flow out of the furnace. Part of the

litharge is absorbed by the hearth which, before starting,
has been lined with bone-ash (now little used) or cement
and lime. During the cupellation the metal surface is completely covered with yellow, glowing litharge which at the
last recedes from the bluish surface of the metallic silver

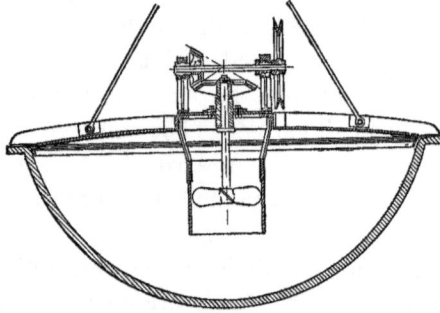

FIG. 51.—Desilverizing Kettle and Auxiliary Apparatus, with stirrer,

(the blick). Since, in spite of the fact that the surface of
the lead is covered with litharge during almost the whole
of the cupelling operation, the oxidation takes place com-

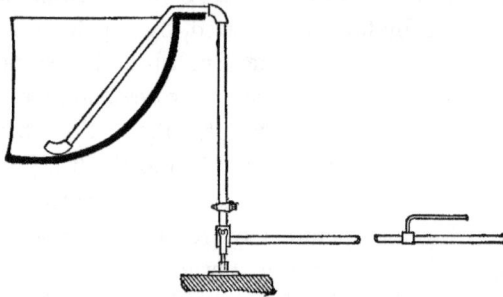

FIG. 52—Desilverizing Kettle and Auxiliary Apparatus, with siphon.

paratively quickly, it must be ascribed to the formation
of a higher oxide of lead, $PbO_2$. In fact, $PbO_2$ is
formed readily in an oxidizing atmosphere at the temperature of the cupelling hearth in the presence of strongly
basic oxides. PbO, however, is itself a strongly basic

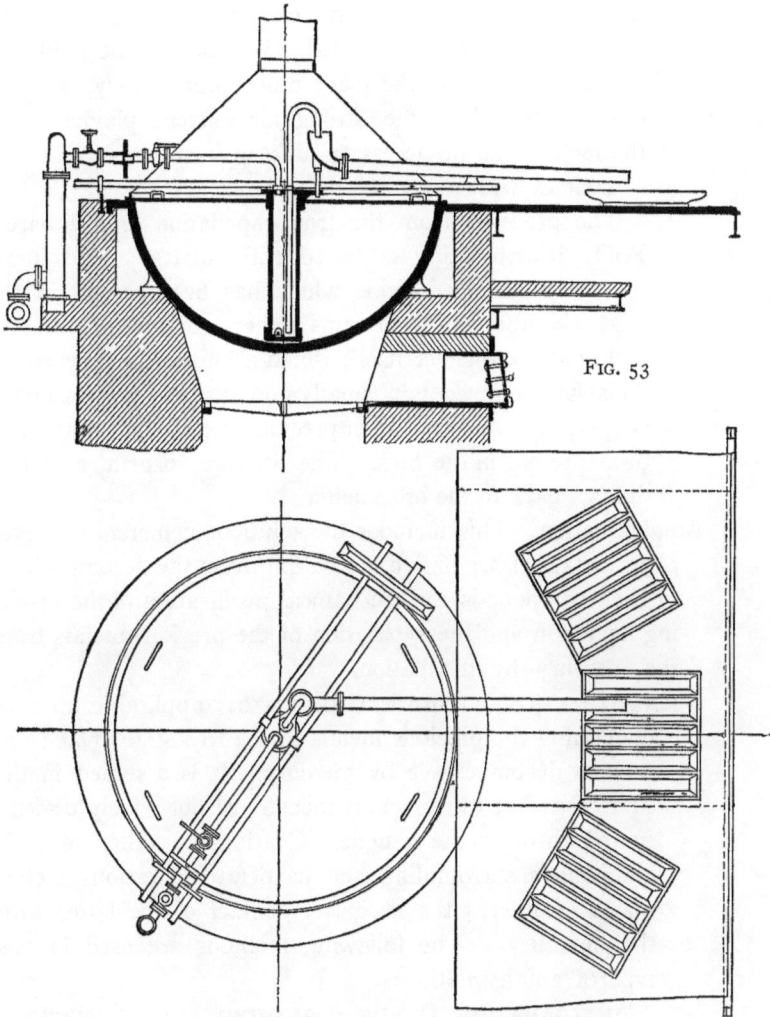

Desilverizing Kettle and Auxiliary Apparatus.
Fig. 51, with stirrer.   Fig. 52, with siphon.   Figs. 53 and 54, with lead pump and molds.

oxide, and can form compounds with $PbO_2$ such as, for example, the lead plumbate found in red lead:

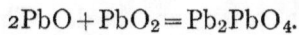

$$2PbO + PbO_2 = Pb_2PbO_4.$$

This compound carries the oxygen of the air through the litharge layer to the lead. Without it, the oxidation of the lead would take place much more slowly, or some device other than the strong air current playing upon the melt would be necessary to keep fresh surfaces of lead in contact with air.

The products from the true cupellation are: litharge, PbO; hearth lining saturated with litharge; lead fume, the volatilized lead oxide which has been caught in the dust chamber and contains other volatile substances; and finally the silver itself, which is known as *blick silver*. This last may contain small amounts of lead, bismuth, copper, etc., also any other precious metals that may have been present in the ores. The litharge, hearth, and flue dust go back to the ore smelter.

**Amalgamation.** This includes the solution in mercury of precious metals, either native or resulting from the decomposition of their compounds, the mechanical purification of the resulting amalgam and the separation of the precious metals from the mercury by distillation.

AMALGAMATION WITHOUT CHEMICALS: applicable to ores containing the precious metals in the free state or as compounds decomposable by mercury. It is assumed in the first case that the precious metals are not widely disseminated through the gangue. Coatings of other metallic compounds surrounding the particles of precious metals may also prevent the necessary contact of the latter with the mercury. The following methods are used in this type of amalgamation:

AMALGAMATION DURING CRUSHING AND SUBSEQUENT MECHANICAL TREATMENT: These methods differ from ordinary ore-dressing processes only by the use of mercury, free or in the form of copper or silver amalgam, in various parts of the apparatus used for crushing and concentrating; in the last case, silver-plated copper plates are used. The apparatus of the ore-

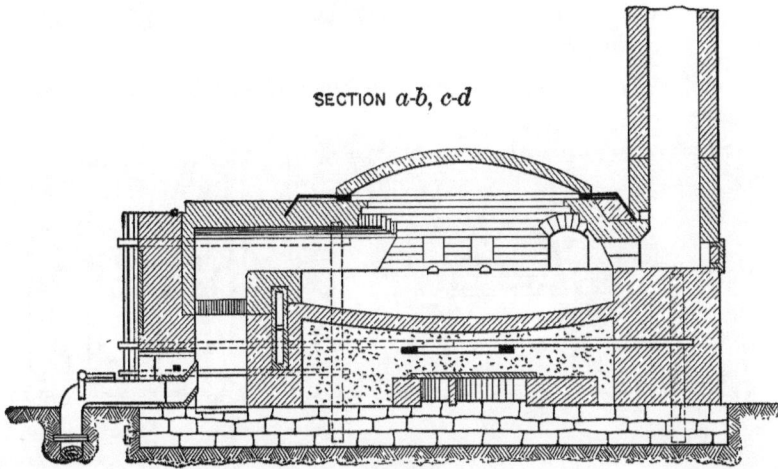

SECTION $a$-$b$, $c$-$d$

Fig. 55

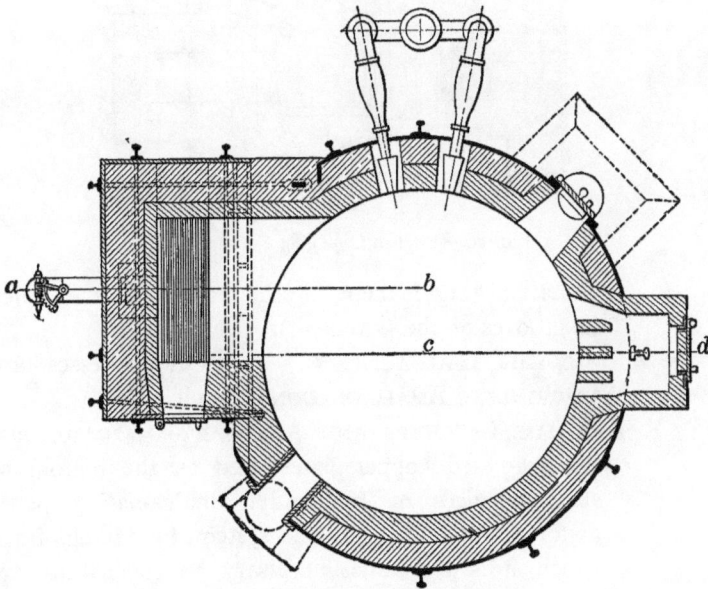

FIG. 56.—German Cupelling Furnace.

dressing plants requires no important changes, so that the following divisions may be made of these amalgamation processes:

HYDRAULIC METHODS in connection with sluice amal-

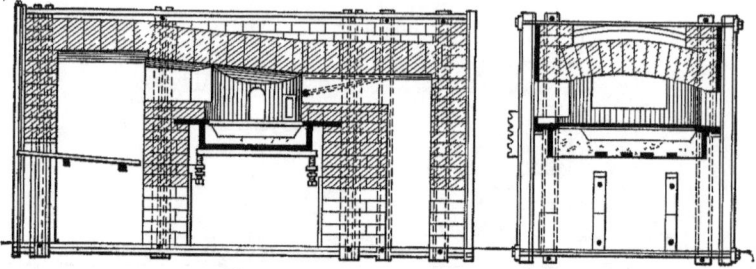

Fig. 57                    Fig. 58

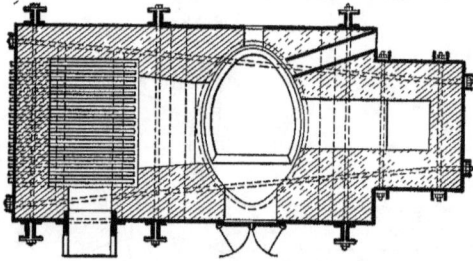

FIG. 59.—English Cupelling Furnace.

gamation (Hg is kept on the bottom of the sluice in grooves of the pavement).

STAMP-MILL AMALGAMATION ⎫ Both processes are
WET-GRINDING AMALGAMATION ⎰ completed on
AMALGAM CATCHERS AND AUXILIARY–AMALGAMATORS.
Amalgamated copper plates used as the bottom of shallow sluices or frequently, amalgamating pans; as for example in the Laszlo system, the tailings from which it is sometimes necessary to concentrate by passing over shaking-tables of the Frue, Wilfley, or Ferraris type, especially for recovering pyrite, arsenopyrite, etc.

CRUSHING AND AMALGAMATION AS SEPARATE PROCESSES.
In this case the amalgamators serve chiefly for mixing

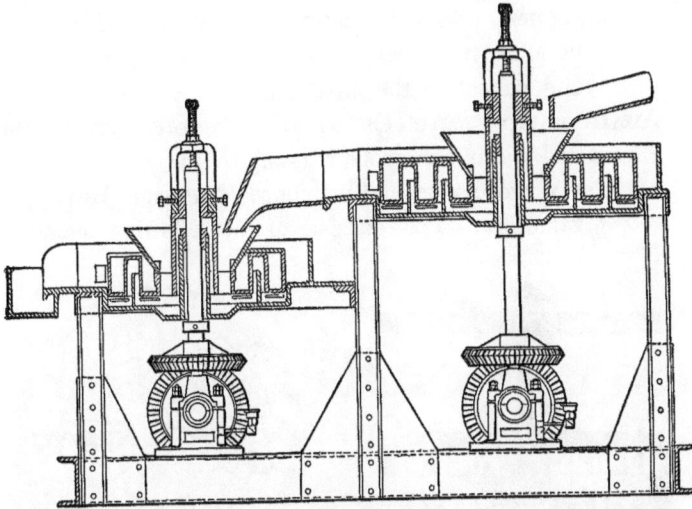

FIG. 60

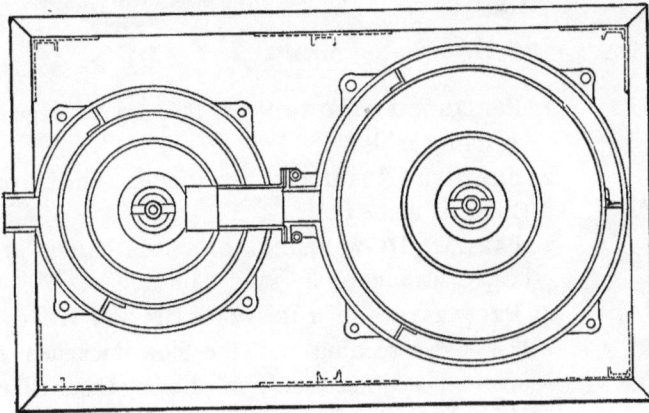

FIG. 61.—Laszlo Amalgamator.

the ore with mercury. They may be classified, according
to their form into

MORTAR AMALGAMATION.

WET-MILL AMALGAMATION OR PAN AMALGAMATION:
Schemnitz amalagmator, Laszlo amalgamator, American amalgamating pans. These contain amalgamated
copper plates as bottoms to shallow sluices, and
various sluices with devices for feeding the mercury
and collecting the amalgam.

AMALGAMATION WITH CHEMICALS: the ores may be amalgamated without previous roasting by the

Patio Process or Old-fashioned, Mexican Heap-Amal-
gamation. This consists of the following operations:

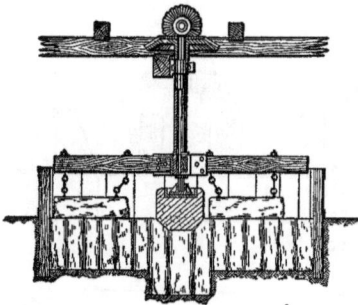 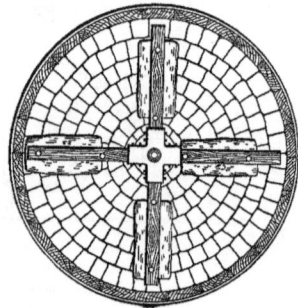

FIG. 62.          Arrastra.          FIG. 63.

1. PRELIMINARY CRUSHING of the ore in edge mills
or stamp mills.
2. FURTHER CRUSHING in wet grinders called *arrastras*
(Figs. 62 and 63).
3. PARTIAL DE-WATERING OF THE SLIMES in low
heaps surrounded by sand dams.
4. PREPARATION OF THE AMALGAMATION HEAPS (*tor-
tas*) by spreading out the now-thickened slime
upon a paved floor surrounded by stones. Diameter
of the floor: 25 to 50 ft., depth 6 to 12 in.
5. INCORPORATION (by treading in with mules) of
from 2 to 5% of common salt (based on the weight
of ore); 0.25 to 0.5% of $CuSO_4.5H_2O$, in the form
of magistral (a product, obtained by the roasting

of pyrite, which contains $CuSO_4$ and $CuCl_2$ besides ferric salts); and 6 to 8 lbs. of Hg per pound of Ag in the ore.

6. WET CONCENTRATION OF THE SLIMES after the completion of the amalgamation.

7. TREATMENT OF THE AMALGAM (see below).

The Patio process, in spite of its great age (already in use 300 years) is a very slightly developed method which gives satisfactory results only with ores that have been prepared by weathering processes, but the mercury loss is always high (200% of the recovered silver). For the chemical reactions see the Kröhnke process.

KRÖHNKE PROCESS: suitable for ores which contain the precious metals either free or in sulphides or chlorides from which the precious metals only are to be extracted. Other metals remain in the amalgamation residue.

REAGENTS: Cuprous chloride dissolved in brine; metals which are electropositive to silver (Zn, Pb, Cu) used in the form of amalgam; and finally mercury.

CHEMICAL REACTIONS: $Cu_2Cl_2$ (or $CuCl$) when dissolved in a concentrated NaCl solution will quickly decompose $Ag_2S$.

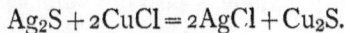

$$Ag_2S + 2CuCl = 2AgCl + Cu_2S.$$

With the degree of fineness practically obtainable in crushing, not over 80% of the silver contained in simple or complex native sulphides is obtained by this method. The stopping of the reaction cannot here be attributed solely to the law of mass action. Opposing the action of the CuCl is the mechanical resistance offered by the $Cu_2S$ crust adhering to the grains of $Ag_2S$. CuCl alone, therefore, is insufficient as a solvent of $Ag_2S$ in amalgamation.

In the presence of metals electropositive to silver, such as Zn, Pb, and Cu, the $Ag_2S$ is reduced to Ag in spite of the coating of other sulphides ($Cu_2S$) or salts. $Cu_2S$ under these circumstances does not hinder the completion of the reaction, but acts as a metallic conductor for the exchange of energy between the more positive metal and silver: it also forms with these and the solution a galvanic couple in which $Ag_2S$ is the cathode and Zn is the anode. If Cu or CuCl is lacking at the beginning of the process, Zn acts upon $Ag_2S$ as follows:

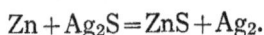

$$Zn + Ag_2S = ZnS + Ag_2.$$

In spite of the seemingly complete separation of Ag from $Ag_2S$, in this case there is another drawback which causes imperfect amalgamation. ZnS, which is insoluble, envelopes the silver and hinders its amalgamation. By using Zn and Hg in a pure NaCl solution, as much as 50% of the Ag that has been converted to the metallic state by the Zn may remain unamalgamated. Copper works more favorably and, in fact, as Kröhnke pointed out, through the formation of CuCl, for CuCl causes the decomposition of $Ag_2S$, but $ZnCl_2$ does not. But, as stated above, CuCl alone is also not sufficient of itself to completely decompose $Ag_2S$, and the solution tension of Cu is weaker than that of Zn. If, now, care is taken at the outset to provide for the presence of CuCl and Zn is added, then the following reaction takes place without the formation of ZnS:

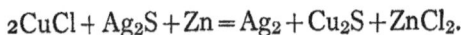

$$2CuCl + Ag_2S + Zn = Ag_2 + Cu_2S + ZnCl_2.$$

$Cu_2S$ does not hinder the amalgamation, as it is more brittle than ZnS and the mechanical movement of the pulp easily exposes the surface of the Ag to the action of the Hg.

OUTLINE OF THE KRÖHNKE PROCESS:

    1. CRUSHING: wet edge-mills called Chilian mills are used.

    2. COLLECTING THE SLIME in a system of settling-tanks,

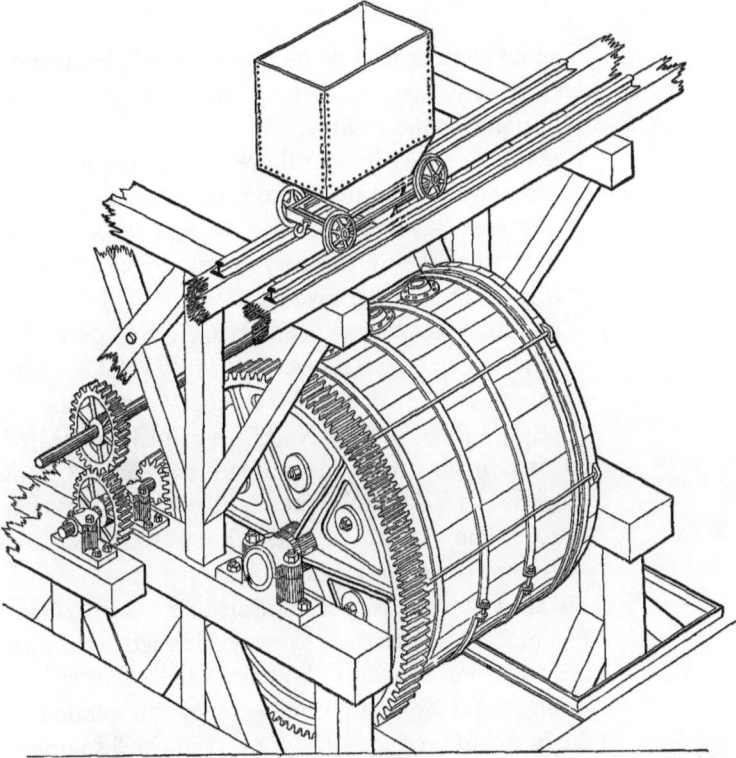

FIG. 64.—Amalgamation Barrel. Length, 6 Feet. Diameter, 5½ Feet. Made of 3-inch Pitch Pine Planks  Wooden (formerly iron) rods pass through the barrel between the heads in order to prevent the charge from balling up. Capacity, 6 tons of ore besides the reagents.

    each of 4000 cu.ft. (100 metric tons) capacity, in which the water is removed by filtration and evaporation.

    3. PREPARATION OF THE AMALGAM: Zn is heated with ten times its weight of Hg under acid water;

Pb with twice its weight of Hg without acid; then filtered.

Zn amalgam contains 14 to 17% Zn.
Pb amalgam contains about 45% Pb.

The Zn content may be increased to 58% by heating and slowly cooling a large mass, in which case filtration is unnecessary.

APPARATUS: cast-iron kettle or, for large quantities, a sheet-metal kettle (bottom of one piece). The bottom edges should project beyond the source of heat so that any mercury which trickles out or is spilled may be recovered.

4. CUPROUS CHLORIDE SOLUTION: 1 part by weight of $CuSO_4.5H_2O + 0.5$ to $1\%$ $H_2SO_4 + 6$ to 20 parts saturated NaCl solutions + scrap copper $(CuSO_4 + 2NaCl + Cu = Na_2SO_4 + 2CuCl)$ are heated by steam for from one-half to two hours (kept saturated with NaCl) in double-walled wooden vats. The space between the walls is packed with a mixture of tar and powdered lime.

5. AMALGAMATION is accomplished in barrels (see Fig. 64). The barrels are charged in succession with the following reagents, boiling NaCl solution, Zn amalgam (or Zn and Hg followed by the rotation of the barrel to produce amalgam), then CuCl solution, followed immediately by the roughly-broken lumps of air-dried slime, which disseminates through the mass, forming a thick paste. The barrels are kept turning slowly (four or five revolutions per min.) for four or five hours. Power required, 8–9 H.P.

6. WORKING UP THE AMALGAM: After completing the amalgamation, the barrels are allowed to stand for from one to three hours; then they are rotated for a short time, at first quickly, then slowly. Every-

thing is then allowed to run out, the barrels are washed out, and the addition of water is continued until its volume amounts to eight times that of the slimes. The thinned slimes are worked up in settling pans, from which the pulp gradually overflows and the amalgam is finally withdrawn at the bottom.

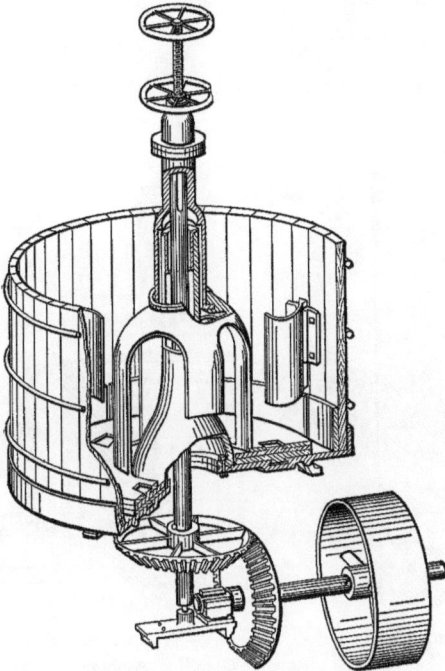

FIG. 65.—American Amalgamating Pan.   FIG. 66.—Amalgam Filter.

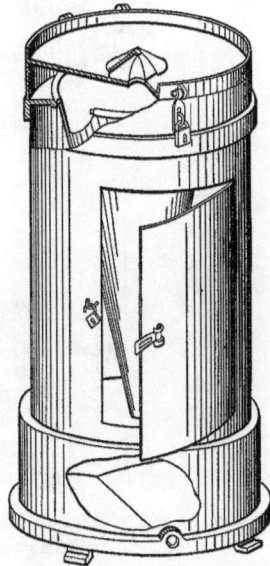

The chemical purification of the amalgam, particularly from copper, is accomplished

Either by treatment with $Ag_2S$ or $AgCl$;

By treatment with $CuCl_2.NaCl$ liquor;

Or by treatment with $NH_3$ and air.

The amalgam is finally filtered (Fig. 66), this being ultimately hastened by a centrifugal machine, and then it is distilled (Figs. 67 and 68).

CASO OR CALDRON PPOCESS:
  REAGENTS: Cu, NaCl, Hg.
  APPARATUS: copper pan.
  APPLICABLE: only for the purer ores carrying native

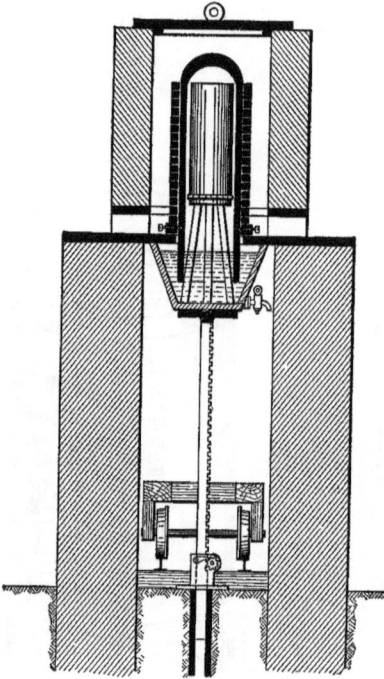

FIG. 67.—The Kröhnke Furnace for
Distilling Amalgam.

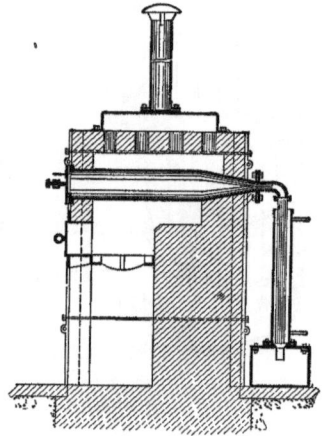

FIG. 68.—Krupp Furnace for
Distilling Amalgam.

silver or AgCl. These are seldom available and the
process is little used at present.

WASHOE PROCESS:
  REAGENTS: $CuSO_4$, NaCl, Fe, and Hg. The principal reagent, as in the Kröhnke process, is CuCl, the
  product obtained by the reaction mentioned above,
  under 4. The action of the reagents is hastened by
  grinding between iron plates in iron pans which are
  heated by steam.

AMALGAMATION APPARATUS: wooden or iron vats with removable bottom plates and mullers (Fig. 65).

PROCEDURE: the ore is crushed in stamp mills, partially de-watered in settling tanks and then subjected to wet grinding, with the above reagents, in steam-heated pans. The pulp is subsequently concentrated and the amalgam purified and distilled.

After a preliminary chloridizing roast, ores may be treated by

Barrel amalgamation.

Pan or vat (Tina) amalgamation.

The preliminary treatment is essentially the same for both processes: Preliminary crushing (rock breakers); drying (usually in rotary cylinders); further crushing with salt (NaCl) in stamp mills (dry stamping) or in other crushing apparatus (edge mills or ball mills). Chloridizing roast in reverberatory furnaces (with stationary hearths and movable stirrers—Pearce, Brown, and Wethey; with movable hearths—Brückner and White-Howell) or in shaft furnaces (Stetefeldt).

The amalgamation proper now takes place either in barrels, in which case there is a preliminary treatment with iron and water to reduce ferric and cupric compounds to ferrous and cuprous compounds and to reduce AgCl to Ag, after which mercury is added; or by pan amalgamation, as in the Washoe process; or by vat amalgamation in wooden vats with copper bottoms and copper mullers. In the latter case the copper acts as the reducing material for cupric and ferric compounds.

The amalgamation procedure is the same as that described above.

## Solution of the Silver in Chemical Solvents.

**Ziervogel Process:** Applicable to matte rich in silver. The matte is concentrated to about 74–75% Cu and carries at the

Mansfeld Works 0.4 to 0.5% Ag. With higher percentages, the Ag segregates in large pellets which is unfavorable for good extraction. The process consists in converting Ag to $Ag_2SO_4$ and the Cu to CuO, leaching out and precipitating the Ag, and smelting the CuO for Cu. The operations are:

1. Breaking of the matte by hand.
2. Grinding in ball mills.
3. Preliminary roasting for two days in a roasting furnace with mechanical stirrers; this converts the copper to $CuSO_4$.
4. Final roasting in a furnace of the same type, which produces CuO and $Ag_2SO_4$.
5. Leaching out the $Ag_2SO_4$ with warm water.
6. Precipitation of the silver with granulated copper in small stoneware vats fitted with filter bottoms.
7. Washing, pressing, drying, and fusing the silver.
8. Re-roasting the leached oxide and further treatment again as under 5, 6, and 7.
9. The $CuSO_4$ resulting from the silver precipitation is concentrated and crystallized to form blue vitriol $(CuSO_4.5H_2O)$.

In the following processes, the Ag is either already present as AgCl in the ore or it is converted into AgCl by a chloridizing roast and leached with chloride or hyposulphite solution from which it is subsequently precipitated. Among these processes, the following have found limited usage:

**The Old Oker-Longmaid-Henderson Process:** leaching of the pyrite, after a chloridizing roast, with water from the condensation tower of the roasters. Precipitation of the Ag with NaI and the Cu with Fe. Recovery of the I by decomposing the AgI with $Na_2S$.

**Augustin Process:** the ore is leached with a saturated brine. One gallon of saturated NaCl solution dissolves at 20°C. 0.169 oz. Ag (100 oz. NaCl: 0.4 oz. AgCl). The Ag is precipitated from the brine by means of Cu.

**Kiss Process:** the ore is leached with $CaS_2O_3$ solution and the silver precipitated by $Ca(SH)_2$.

**Russell Process:** leaching with $4Na_2S_2O_3 \cdot 3Cu_2S_2O_3$ solution and precipitation with $Na_2S$.

**Patera Process:** this process is used extensively, especially in plants designed and built by Ottaker Hofmann. The Patera-Hofmann process consists of the following operations:

1. A PRELIMINARY LEACHING with water to remove chlorides of the base metals.

   APPARATUS: wooden vats, 15–20 ft. in diameter and 5 ft. deep with removable filter bottoms of lattice work. All wooden surfaces are coated with asphaltum.

2. LEACHING OUT THE SILVER with sodium "hyposulphite" containing $0.25$–$0.5\%$ $Na_2S_2O_3$ $5H_2O$. — $100$ g. of $Na_2S_2O_3 \cdot 5H_2O$ will dissolve $40$ g. of $AgCl$, forming $Ag_2S_2O_3 \cdot Na_2S_2O_3 \cdot 2H_2O$. Same apparatus as in 1.

3. PRECIPITATING of the Ag with $Na_2S$ or calcium polysulphide, the latter being prepared by boiling milk of lime with sulphur. The calcium polysulphide solution is preferred because it prevents the concentration of $Na_2SO_4$ in the solutions, the $SO_4$ being precipitated as $CaSO_4$. The $Na_2S$ resulting is oxidized to $Na_2S_2O_3$, so that no fresh supply of this salt is required.

4. FILTRATION: the precipitate is transferred to a settling tank and from there to a filter press.

5. DRYING AND ROASTING in a reverberatory furnace.

6. CHARGING INTO A HOT LEAD BATH and cupelling the rich argentiferous lead.

**Nitric Acid Parting:** This process, known as inquartation, is only used to-day, except for Au-Ag parting in assaying, according as the demand for the resulting $AgNO_3$ permits. Ag is dissolved by $HNO_3$, leaving the Au unattacked. If the ratio of Au to Ag in the alloy is between $1:3$ (hence the name inquartation) and $1:1.75$, the Au remains behind in a coherent state after heating the alloy with $HNO_3$. If the Au content is higher, the resulting Au carries Ag even after repeated treatment with $HNO_3$. With less Au,

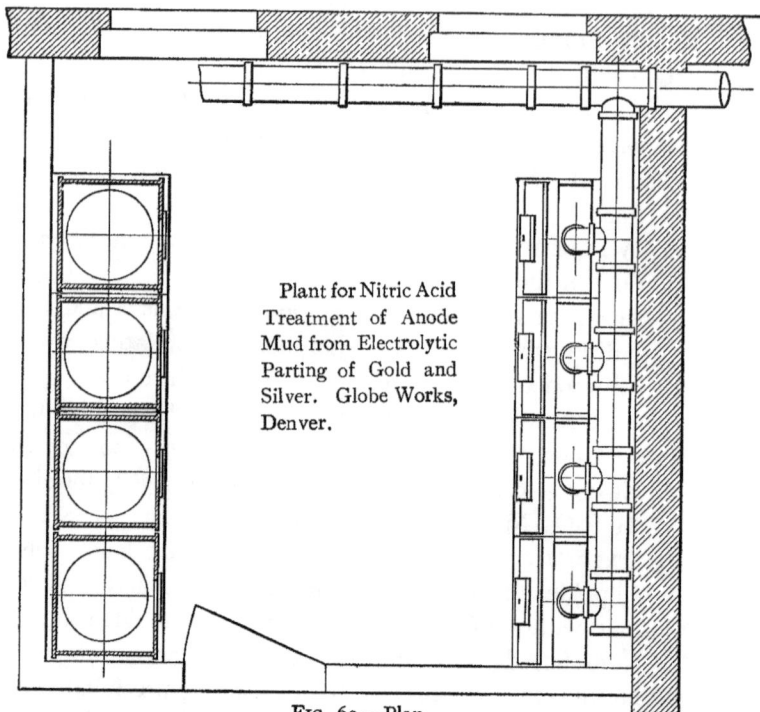

Plant for Nitric Acid
Treatment of Anode
Mud from Electrolytic
Parting of Gold and
Silver. Globe Works,
Denver.

FIG. 69.—Plan.

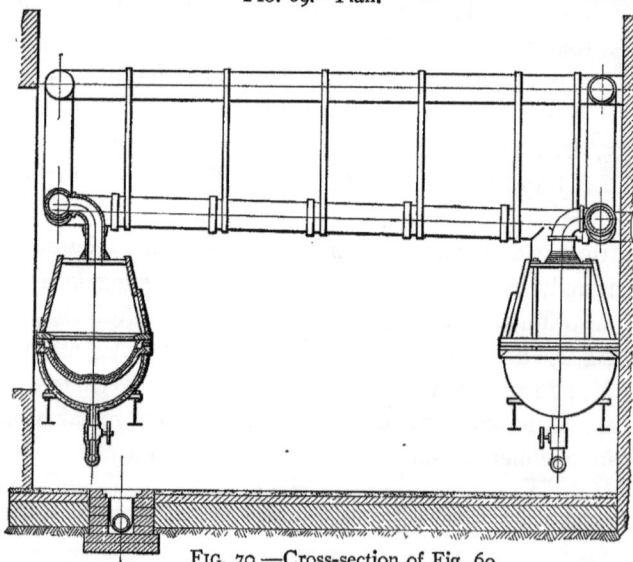

FIG. 70.—Cross-section of Fig. 69.

the Au remains behind as a powder.  Apparatus necessary: glass, porcelain or stoneware vessels, less frequently platinum dishes.  An apparatus for parting Au-bearing anode mud by the $HNO_3$ method is shown in Figs. 69–70.

**Sulphuric Acid Parting:** applicable for Au-Ag alloys which contain not more than 1 oz. Au to 2–4 oz. Ag and less than 10% Cu.  The $H_2SO_4$ dissolves the Ag and Cu, forming sulphates:

$$Ag_2 + 2H_2SO_4 = Ag_2SO_4 + 2H_2O + SO_2,$$

if the granulated alloy is boiled with an excess of $H_2SO_4$.  On a small scale, porcelain vessels are used, on a large scale, cast iron.  The acid solution is diluted with water in lead-lined vats and clarified from small particles of gold thereupon the silver is precipitated on scrap copper or iron in other lead-lined vats.  The precipitated silver is removed and compressed.  For further treatment, see Silver Refining.  For treatment of the gold, see Gold Refining (page 21).

**Chloridizing Fusion:** conducting Cl into the melted Au-Ag alloy in order to convert the Ag into AgCl is no longer used.

**Electrolysis** of Ag-Au-Cu alloys, see Copper.

**Solution of the Base Metals:** In these methods the silver remains in the residue.  According to the nature of the ore and the condition of working, one of the following processes is used:

**Freiberg Vitriolization Process:** adapted to roasted matte.  In contrast to the Ziervogel process the ore is so roasted that the Cu is converted to CuO and Ag to the metallic state.  The CuO is dissolved as sulphate by treatment with $H_2SO_4$.  The silver-bearing residue is smelted with silver-lead ores and the solution is worked up into blue vitrol ($CuSO_4 \cdot 5H_2O$).

**Rössler's Sulphurizing Fusion.** This consists in melting Au-Ag-Cu alloys with S.  There remains an Au-Ag alloy,

Electrolysis Möbius Process.

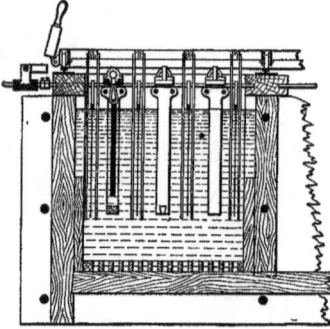

FIG. 71.—Cross-section of Electrolytic Cell.

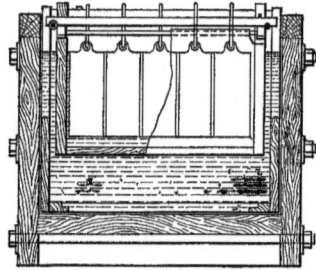

FIG. 72.—Anode Cell with Five Anode Plates of Doré Silver.

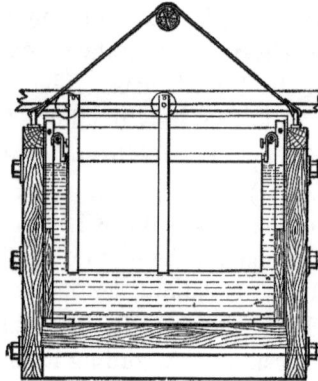

FIG. 73.—Cathode and Supporting Device.

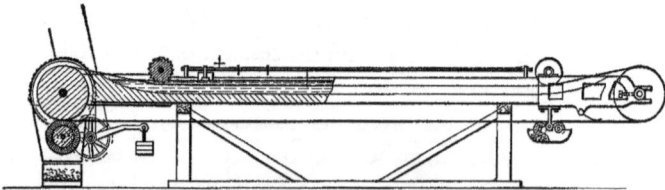

FIG. 74.—Modern Möbius Apparatus. Cathode, endless belt of sheet silver having below devices for greasing and brushing. This belt is drawn through a shallow vat. Anodes, bars of doré silver placed in shallow troughs having porous bottoms.

which is parted either by the sulphuric acid method (see above) or by electrolysis (see Silver Refining). The $Cu_2S$, which contains some $Ag_2S$, is converted to argentiferous copper by an oxidizing smelt and then electrolyzed.

**Hartz Vitriolization Process.** This process converts precious metals carrying copper into blue vitriol by the action of air and dilute sulphuric acid on the granulated alloy:

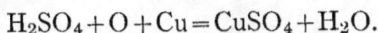

$$H_2SO_4 + O + Cu = CuSO_4 + H_2O.$$

The solution is clarified and converted by concentration into blue vitriol. The slime carrying the precious metal, after being washed and dried, is refined by cupellation.

**Electrolytic Refining:** see Copper.

**Electrolytic Refining of Zinc Skimmings:** see Zinc.

## (B) Refining Silver

The aim in silver refining is to remove the last traces of impurities and to effect also the separation of the silver from other precious metals. It consists usually of an oxidizing fusion in which oxidizing fluxes are utilized as well as the atmospheric oxygen, and, if other precious metals are present, of electrolysis.

1. **Fire Refining** of the crude silver either by melting under airblast, i.e., carrying out cupellation in small reverberatory furnaces with oxidizing flame, or

FUSING WITH NITER, which is common in working up precipitated silver (cement silver) which carries small quantities of Fe or Cu (apparatus: crucibles), or by

FUSING WITH $Ag_2SO_4$ (Rössler's method), adapted to Ag carrying Bi. This process is also carried out in crucibles.

2. **Electrolytic Refining.** According to Möbius and Wohlwill, if gold-bearing silver (the so-called *doré silver*) in the form of small plates or bars, is used as the *anode* in an aqueous solution of $HNO_3$ (1%), $AgNO_3$ (0.5% Ag) and $Cu(NO_3)_2$ (4% Cu), the Ag will be dissolved and deposited upon a sheet of pure Ag, which serves as the *cathode*.

The current density recommended is 30 to 20 amperes per sq.ft. The E.M.F. is 1.5–1.4 volts. The apparatus is shown in Figs. 71, 72, 73, and 74, page 52.

## Properties of Refined Silver:

SPECIFIC GRAVITY: 10.5.

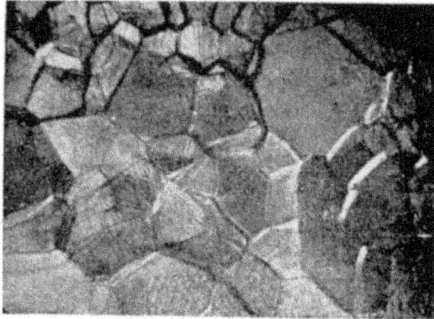

FIG. 75.

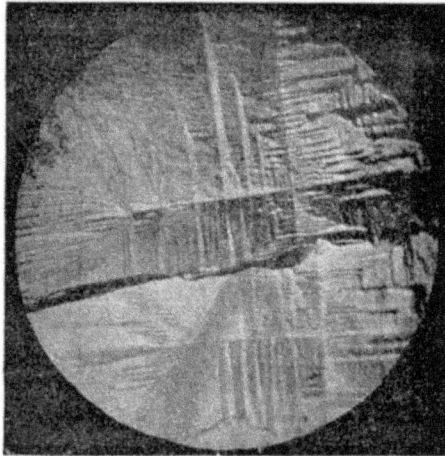

FIG. 76.—Fused Silver.   Dendritic Surface ($\times 33$).

COLOR: white with brilliant lustre.
DUCTILITY: tough, very ductile.
MELTING-POINT: 961° C. (1762° F.).
VAPORIZED readily at 1200–1500° C. (2192–2732° F.).

THERMAL AND ELECTRICAL CONDUCTIVITY best of all metals
ALLOYS WITH MOST METALS: the most important metallurgical alloys are those with Au, Pb, Hg, Cu and Zn.

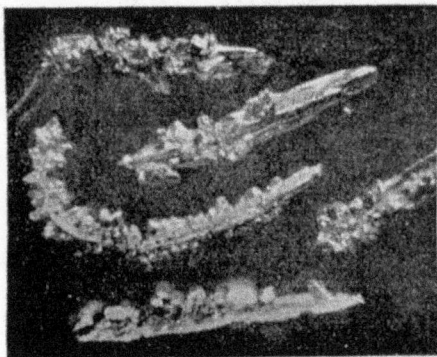

FIG. 77 —Electrolytic Silver (Mag. 16).

CHEMICAL BEHAVIOR: not very active, although its
solution tension is greater than that of gold. Silver
compounds are easily dissociated and are sensitive to light
(photography). As solvents for metallic Ag, concentrated
$HNO_3$ and $H_2SO_4$ are used. For AgCl other metallic
chlorides and " hyposulphite " are employed.

# MERCURY

## Sources

**Natural Sources:**

FREE, usually containing Ag and less frequently Au.

As CHLORIDE, HgCl, calomel (rare).

As SULPHIDE, HgS, cinnabar, the most important ore of mercury. It is finely disseminated through the gangue and the per cent of mercury present is usually low. With other sulphides in tetrahedrite.

**Other Sources:**

AMALGAMS resulting from the use of mercury in extracting other metals or from contact with metals in other ways.

RESIDUES and unmarketable products produced in making mercury pigments (HgO and HgS).

## (A) Extraction of Mercury

**Roasting Accompanied by Distillation:**

**Oxidizing Roast** is the method commonly used in treating mercury ores. Since the dissociation temperature of HgO is very low (400° C.), Hg results directly from roasting at high temperatures:

$$HgS + O_2 = Hg + SO_2.$$

APPARATUS: Shaft furnaces for lump or briquetted ores. (Figs. 78–79.)

SELF-FEEDING ROASTERS for finely-divided ores. (Figs. 80–83.)

REVERBERATORY FURNACES for easily reducible or sintering ores. To all of these furnaces, a system of iron or clay

CONDENSATION TUBES is attached, these tubes being partly cooled by air, and partly by water.

The products of these processes are:

LIQUID MERCURY, which collects under water in the lowest part of the condenser.

MERCURIAL DUST or SOOT, made up of flue dust from the

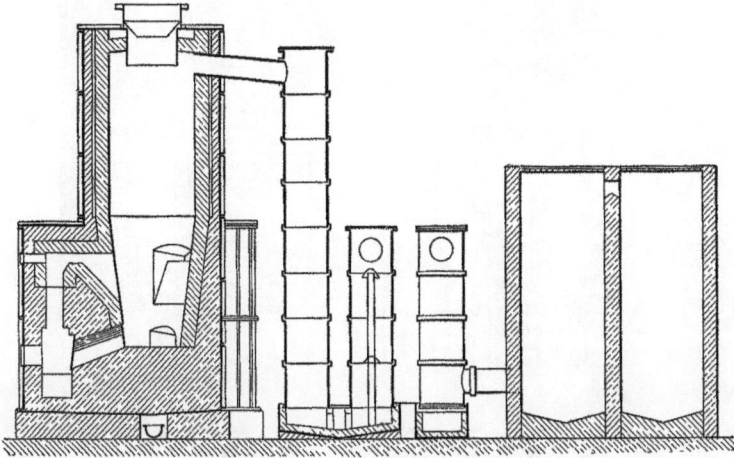

FIG 78.

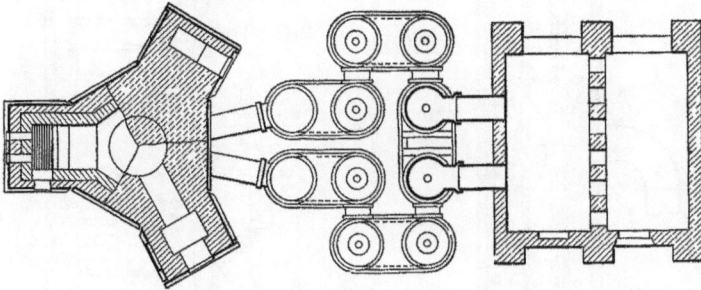

FIG. 79.

ore and sublimated mercury compounds (HgO, HgS, HgSO$_4$) which collect on the walls of the condenser and mechanically retain globules of mercury.

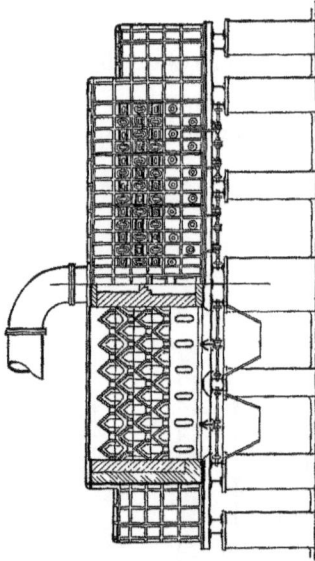

Fig. 81.

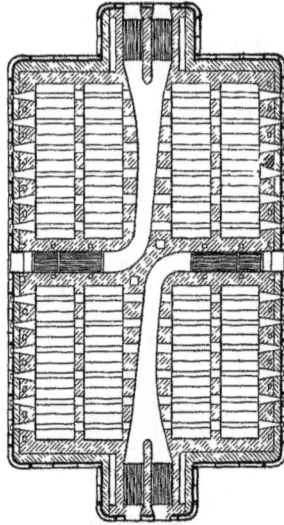

Fig. 83.

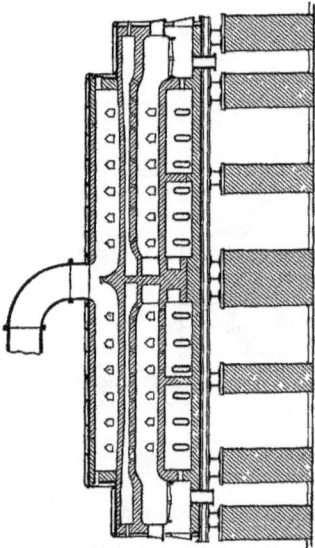

Fig. 80.

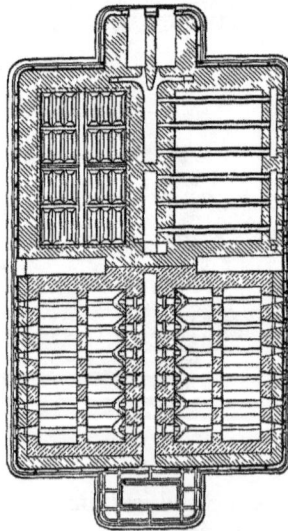

Fig. 82.

**Heating with Desulphurizing Fluxes** (Fe and CaO), is used only for especially rich ores, or for residues from the manufacture of cinnabar. For this purpose retorts are used such as have been already described under Amalgam Distillation (see Silver).

# (B) Mercury Refining

**Filtration** of the liquid mercury that collects in the condensing tubes is frequently sufficient. The use of presses is necessary in working up the " dust " or " soot."

**Distillation** is used for amalgam (see Silver).

**Washing with Acids** is employed when the mercury is to be used for scientific purposes.

**Properties of Mercury:**

SPECIFIC GRAVITY: 13.5.

COLOR: bluish-white.

TENACITY: liquid at ordinary temperatures.

FREEZING-POINT: $-39.4°$ C. $(-38.9°$ F.).

VOLATILIZES slowly at ordinary temperatures.

BOILING-POINT: $360°$ C. $(680°$ F).

ELECTRICAL CONDUCTIVITY: about 0.017 that of Ag.

ALLOYS: form readily with Au, Ag, Cu, Pb, Sn, Cd, Zn, with the alkalies and alkaline earths. More difficultly with Ni, Co, Fe, Mn, Cr, W and the earths.

CHEMICAL BEHAVIOR: The affinity of Hg for the halogens and for S is great; slight for oxygen. The chlorides HgCl and $HgCl_2$, as well as the sulphide HgS, may be sublimed without decomposition. HgO first forms at about $300°$ C., but dissociates very readily at $400°$. The best solvents for Hg are $HNO_3$ and aqua regia. Hg may form a univalent as well as a bivalent cation in solution.

# COPPER

## Sources

**Natural Sources:**

NATIVE COPPER: surrounded or accompanied by sulphide or oxide copper ores, clay and shale.

RED COPPER ORE, cuprite, $Cu_2O$, carrying 88.8% Cu. Associated with sulphide copper ores, spathic iron ore, clay or shale, and earth.

BLACK COPPER ORE, thenorite, CuO (rare).

COPPER GLANCE, $Cu_2S$, free but not widely distributed.

COPPER PYRITES, chalcopyrite, $Cu_2S.FeS.FeS_2$. This, the most important copper ore, contains 34.6% Cu. It is associated with galena, blende, shale and slate.

PEACOCK ORE, bornite, $(Cu_2S)_3.FeS.FeS_2$. Not so abundant as chalcopyrite.

BLUE VITRIOL, $CuSO_4.5H_2O$. Formed by the weathering of sulphide ores and associated with their gangue.

MALACHITE, or green carbonate of copper; $HO-Cu-CO_3-Cu-OH$; formed by the weathering of other copper minerals. Associated with similar minerals and in similar localities.

AZURITE, or blue carbonate of copper; $HO-Cu-CO_3-Cu-CO_3-Cu-OH$; associated with malachite.

**Other Sources:**

MATTE containing copper, speiss, slag and alloys from lead and nickel smelting.

SCRAP METAL and residues from metal-working processes.

## (A) Enriching and Concentrating Processes

The percentages of Cu, as given above, refer to the pure mineral, free from gangue. By means of the gangue, however, the Cu content of most ore bodies is reduced to a few per cent. From these ores the copper can neither be successfully leached out by chemical means nor can it be sufficiently concentrated by mechanical processes so that the ores can be successfully fused directly for metal, as this would be so impure that refining would be out of the question. It is necessary, therefore, to effect a chemical concentration by a fusion. Of all the metals that come into consideration here, copper has the greatest solution tension in fused sulphides; it will, therefore, withdraw the sulphur from other metals in so far as the total amount of sulphur present in a fused mixture of sulphides and other compounds is insufficient to form the lower sulphides of all the metals. Upon this principle are based the following

### Roasting and Smelting Operations:

1. **An Oxidizing Roast** effects the removal of the sulphur from sulphide ores, leaving, however, more than

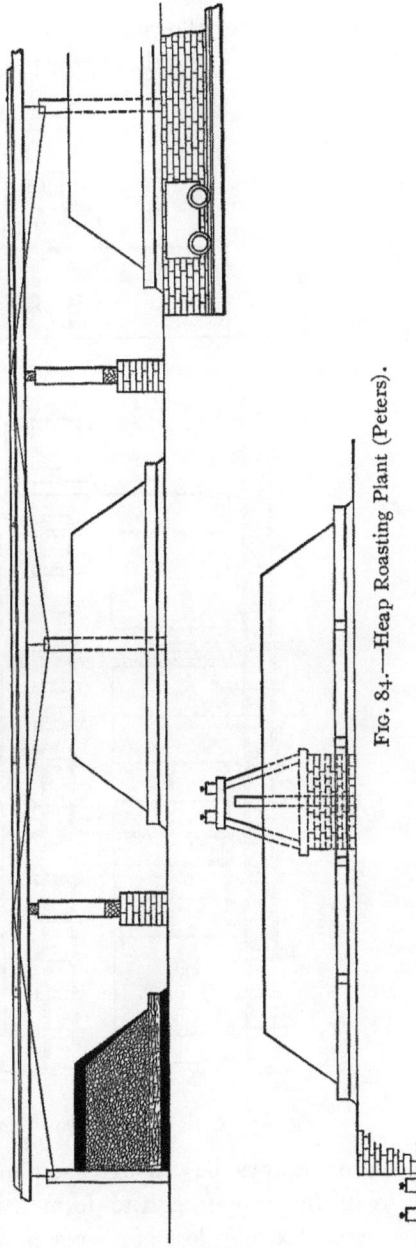

FIG. 84.—Heap Roasting Plant (Peters).

enough, after allowing for a loss due to reaction in the furnace, to form $Cu_2S$ with the copper present. In matte smelting, there

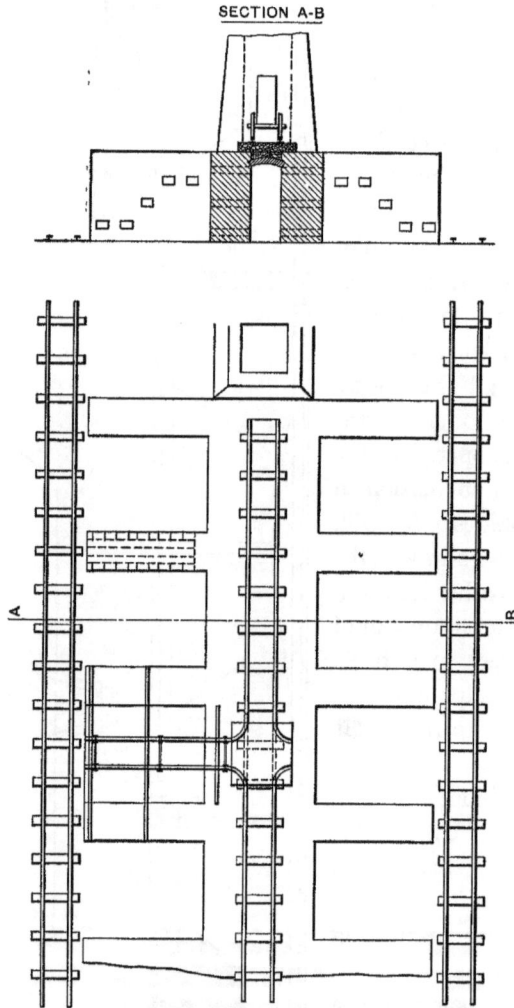

FIG. 85.—Stall Roasting Plant (Peters). Scale, 1 : 150.

must always be enough sulphur present to form $Cu_2S$ with the copper and to form a certain amount of FeS, to prevent a high loss of copper in the slag; the FeS and $Cu_2S$

should be present in such quantities as to correspond to the formula $FeS.Cu_2S$.

In crushing the ore, which is usually in the form of large lumps to make it suitable for roasting, it should be observed that the finer the ore the greater the cost of roasting, the greater the loss in flue-dust, and the greater the hindrance in subsequent leaching operations. The size of grain desirable in roasting is from 6 to 60 mm. In choosing a method of crushing, one that is quick and gives the smallest amount of fines should be selected. Hand breaking has advantages, for the fines are then under 10% as a rule. Machines which produce the smallest quantity of fines usually require the most power. If the nature of the ore requires that it be reduced to a small grain, rolls, stamp mills and ball mills are used. The following processes and and apparatus may be mentioned.

HEAP ROASTING: Open heaps of ore having a trapezoidal or hemispherical cross-section are constructed upon some combustible material in order to start the roasting. (See Fig. 84.)

STALL ROASTING: To serve as a protection against the weather and to prevent damage to the neighborhood, walls may be built to protect the heap. These walls at first consisted of board fences, earth banks, and finally stone walls, from which came the so-called "stall." Stall roasting is essentially the same as heap roasting except for the protective walls, which also serve to regulate the amount of air. The best stall construction is Peters' modification of the Swedish stalls. (See Fig. 85.)

PYRITE BURNERS: these may be regarded as a further step in stall construction. Many different forms have been developed, among which may be mentioned: kilns (Fig. 86), lump pyrites burner (Fig. 87), fine pyrites burner, Gerstenhöfer's fine pyrites furnace, Hasenclever-Helbig furnace, Olivier-Perret furnace and the Maletra-Shaffner roasting furnace (Fig. 88).

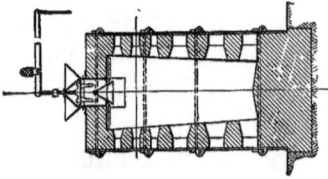

FIG. 86.—Mansfeld Kiln for Matte Roasting. Scale, 1 : 150.

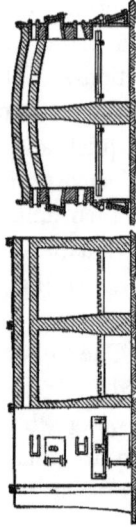

FIG. 87.—Lump Pyrite Furnace (Lunge). Scale, 1 : 150.

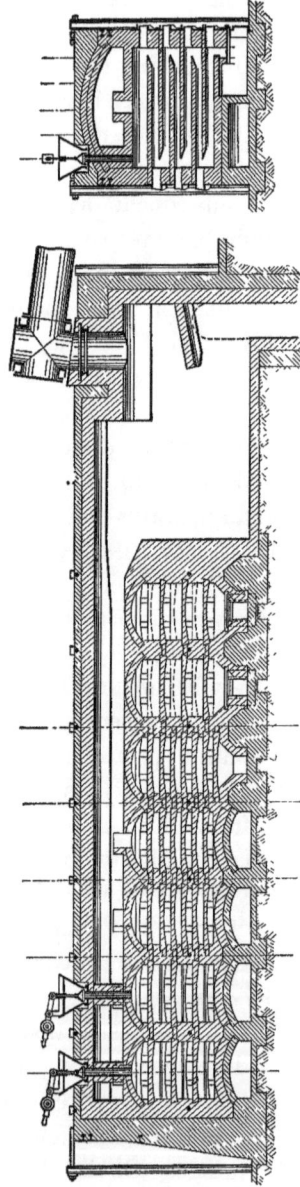

FIG. 88.—Maletra-Schaffner Furnace (Lunge). Scale, 1 : 150.

REVERBERATORY ROASTING. As compared with other reverberating furnaces, the following points are character-

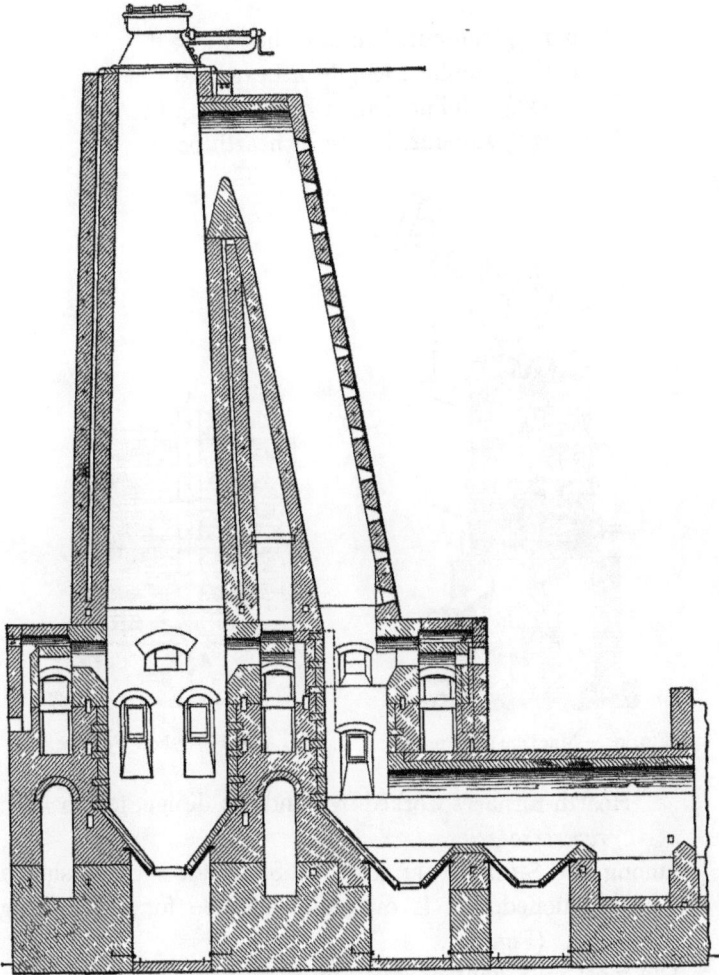

FIG. 89.—Stetefeldt Shaft Furnace.

istic: They should combine a small firebox with a relatively long hearth, and the length of the latter depends upon the amount of heat evolved from the roasting ore. Ores carrying 10% or less of sulphur are seldom roasted in reverberatories. The relation between length of hearth and per cent of sulphur is as follows:

With 10% sulphur, length of hearth 15 ft.
With 15% sulphur, length of hearth 30 ft.
With 20% sulphur, length of hearth 45 ft.
With 25% sulphur, length of hearth 60 ft.

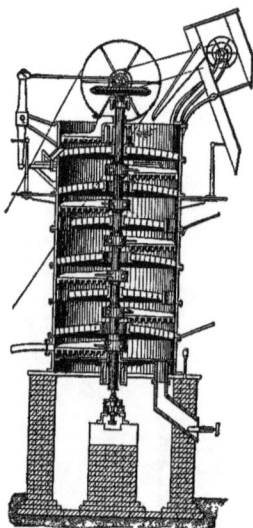

FIG. 90.—MacDougall Furnace.

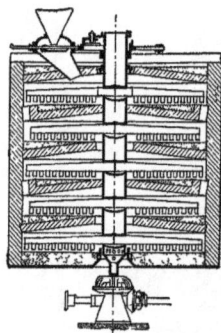

FIG. 91.—Herreshoff Furnace.

Hearth-furnaces worked by hand are designated as *hand reverberatories.*

Among the SHAFT FURNACES the Stetefeldt furnace should be mentioned. It is especially suitable for chloridizing roasting. (Fig. 89).

The most important reverberatories with mechanical stirrers are the following:

PARKES FURNACE: this has two circular superimposed hearths with radial stirring arms attached to a central shaft. The firebox is beside the lower hearth.

MACDOUGALL FURNACE (Fig. 90), Herreshoff furnace (Fig. 91) and the furnaces of Humboldt and Kaufmann: these have five to seven circular hearths placed one above the other. Horizontal arms, extending from a vertical shaft, work the ore in such a way that on the upper hearth it is slowly moved toward the periphery. Here it falls to the next hearth, is worked toward the centre, falls to the third hearth and so on to the last. The arms are easily removed and replaced by new ones.

O'HARA FURNACE: In its original construction this furnace consisted of two superimposed hearths through which rakes resembling ploughshares were drawn by means of an endless chain. The latter was run upon sprocket wheels. During interruptions in the process, the chains should be dropped into a groove in the hearth in order to protect them from the hot gases. This apparatus is untrustworthy and suffers greatly from hot gases.

ALLEN FURNACE: Allen obviated the above difficulties by placing tracks in his furnace upon which carriages carrying the rabbles were supported. The carriages were drawn by the chain as before.

BROWN FURNACE: Brown placed the tracks and supporting carriages in compartments built in the side walls of the furnace. Brown's furnace is also constructed in the form of a broken circle, the so-called elliptical furnace, with external fireboxes. (Fig. 92.)

HIXON FURNACE: Hixon finally placed the supporting trucks and the dragging device (iron cables instead of chains) in a channel built in the bottom of the furnace.

HOLTHOFF-WETHEY FURNACE: the rails and trucks for the tube-like rabble holders were placed entirely outside the walls of the furnace. The space beneath the hearth served mainly as a cooling hearth.

PEARCE FURNACE (see Gold, page 3, Figs. 1 and 2): The hearth is circular. The rabbles are held by tube-like arms through which air for roasting is supplied. The

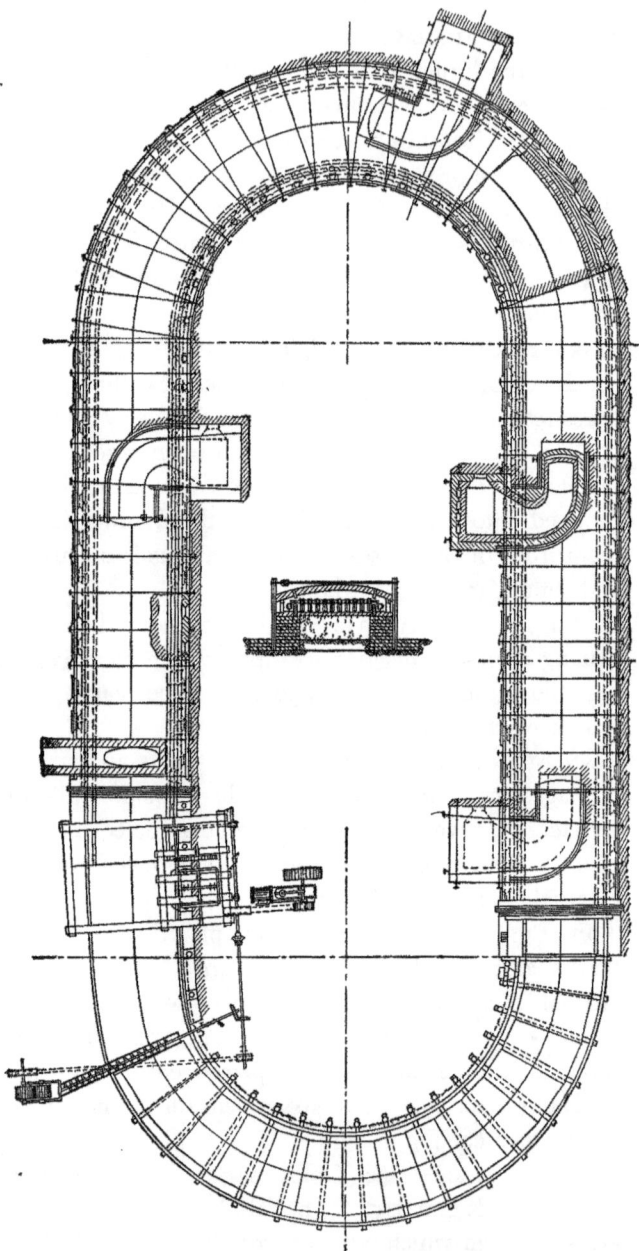

Fig. 92.—O'Hara-Brown Elliptical Furnace.

rabble arms are driven from a central shaft and project through a slot in the inner wall of the furnace. The slot is kept closed by an iron ring which moves with the rabbles. The Pearce furnace is built both with a single hearth and with two hearths one above the other. Auxiliary heating, if necessary, is supplied by external fire boxes. The space under the roasting hearth serves as a dust chamber.

WETHEY FURNACE: This is a furnace for fine pyrite and , consists of four superimposed hearths. The rabbles make a circuit through two hearths. Two of these furnaces

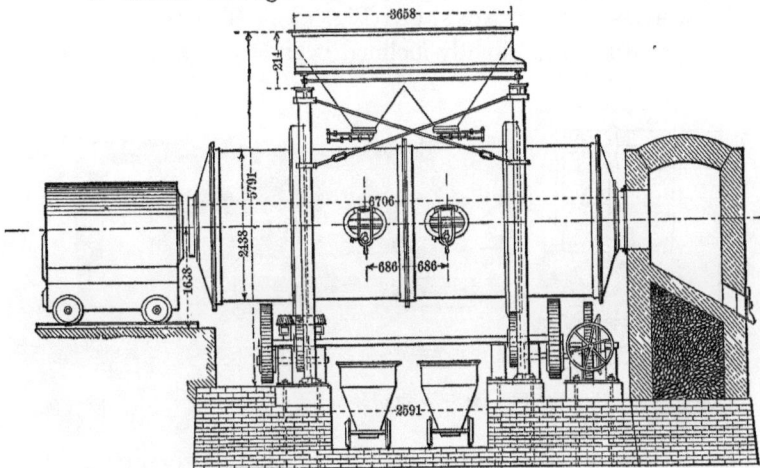

FIG. 93.—Bruckner Furnace.   Scale 1:150.

may be placed side by side and the rabbles moved by the same machinery.

SPENCE FURNACE: This is a furnace for treating fine pyrite. It consists of four superimposed hearths, the rabbles being drawn back and forth over the hearths by means of horizontal rods.

KELLER FURNACE: This is a furnace for fine pyrite. It has five superimposed hearths and its operation is similar to that of the Spence furnace. Two furnaces are usually placed side by side and operated by the same engine.

Among furnaces with moving hearths the following have found favor.

BRÜCKNER REVOLVING FURNACE (Fig. 93): This furnace, which is widely used, consists of a cylinder resting upon friction wheels to which power is applied by means of toothed gears. The heated gases from an external fireplace pass axially through the cylinder. At the opposite end are the dust chamber and stack. The cylinder is charged from hoppers placed above manholes in the furnace walls, the holes being subsequently closed. After the roasting is completed the furnace is discharged through the same holes into cars.

WHITE-HOWELL FURNACE (Fig. 94): The roasting hearth consists of a slightly-inclined cylinder, open at both ends.

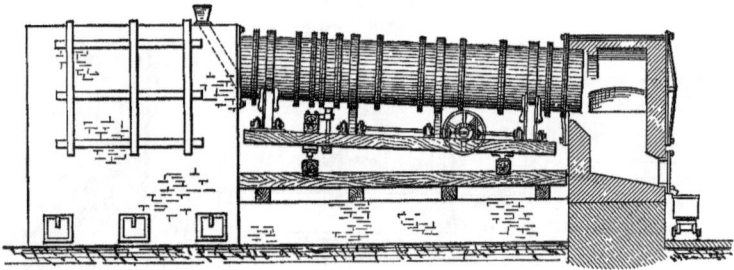

FIG. 94.—White-Howell Furnace.

It is made of sheet iron and lined with firebrick. The furnace is fed through a tube at the higher end and is discharged into a walled chamber at the lower end. Beside this chamber is the fireplace from which the hot gases pass over the roasting ore. At the higher end is a dust chamber from which leads the flue to the stack.

HOCKING-OXLAND FURNACE: This is similar in construction to the White-Howell, but somewhat differently built in the arrangement of the firebox and the chamber for collecting the roasted ore.

ARGALL FURNACE (Fig. 95): Within a broad cylinder are enclosed four smaller, somewhat shorter cylinders slightly inclined. The large cylinder is fitted with firebox and dust chamber similar to those previously described. The ore is fed into each narrow cylinder in turn as it reaches its highest point during the revolution of the broad cylinder, and is discharged at the opposite end, which lies next to the firebox. A chamber below the large cylinder receives the ore.

MUFFLE FURNACES: These are used for the chloridizing roasting of copper ores, especially roasted pryite carrying copper. The construction, shown in Fig. 96, page 72, has four iron charging tubes with gates for closing. The charge is worked by hand through doors in the side of the muffle.

FIG. 95.—Argall Furnace. Scale, 1 : 150.

The gases are conducted away through an iron tube near the firebox. In order to lessen the danger of overheating the charge, the muffle arch is made of double thickness.

2. **Smelting for Matte** consists in uniting the copper of the roasted product with part of the sulphur, uniting the rest of the sulphur with part of the iron and slagging the remaining iron together with other materials. In order to prevent high slag losses, it is best not to attempt too great a concentration of copper. A matte carrying 50% copper should be regarded as the maximum and in most cases a lower grade is produced. The concentration of the ore to matte obtained by these processes fluctuates widely, viz., between 12:1 and 3:1 or, in other words, between 12 and 3 tons of ore yield 1 ton

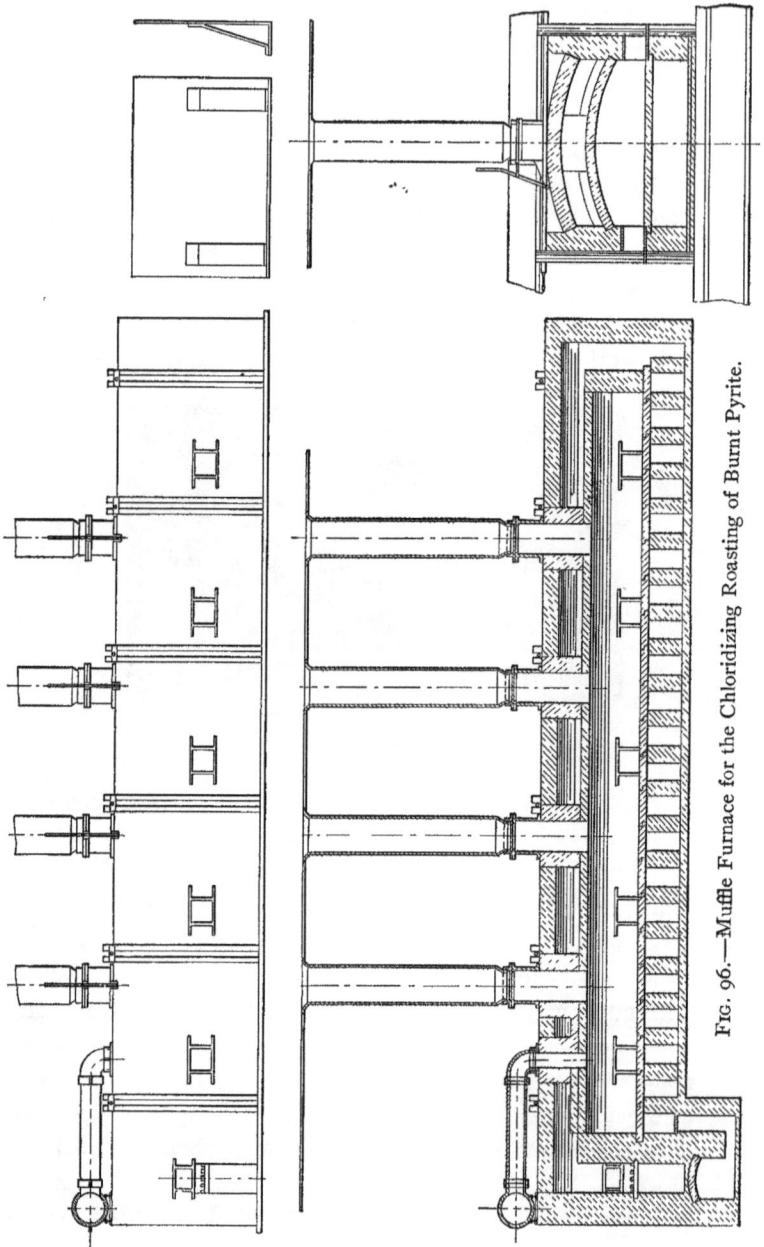

FIG. 96.—Muffle Furnace for the Chloridizing Roasting of Burnt Pyrite.

of matte by the process of roasting and smelting. In roasting attention must be paid to these limits.

The roasted product contains a mixture of the oxides, sulphides, sulphates, arsenates, antimonates, and silicates of the various metals contained in the ore.

PRINCIPLES OF THE SLAG CALCULATION: The slag should lie between a singulo- and a bi-silicate, preferably in the vicinity of a singulo-silicate with FeO as the predominating base. Exceptions: With high iron and zinc content: between a singulo and a sub-silicate. With high $SiO_2$ content: between a bi-silicate and a tri-silicate. It should be borne in mind that the furnace temperature must be raised if high silica is used, thus increasing the danger of reducing the iron compounds:

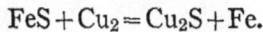

$$FeS + Cu_2 = Cu_2S + Fe.$$

APPARATUS FOR MATTE SMELTING:

1. SHAFT FURNACES: From the original narrow shafts with rectangular or trapezoidal cross-sections, a change has been made in favor of wider furnaces with circular, oval, or long rectangular cross-section, and recently to six-, seven- and eight-sided sections.

The height of the shaft depends upon the iron content of the ore and the nature of the fuel. The height diminishes as the iron content increases. Furnaces using charcoal should be made higher than those using coke. High zinc necessitates a low furnace, as otherwise deposits of zinc oxide will be formed.

The chemical action of the matte and slag have led to the adoption of cooled furnace walls. Whereas in lead smelting only a relatively small portion of the wall is cooled, in copper smelting not only the whole bosh but in some cases the entire shaft is surrounded by a water-jacket. For these relatively large cooling plates, cast iron should be avoided as much as possible. Besides wrought iron, copper is

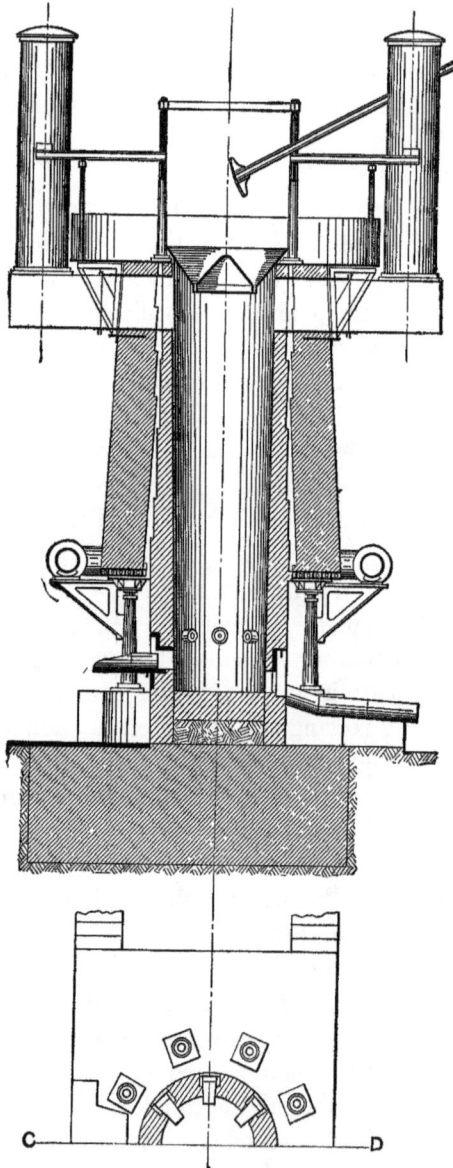

FIG 97.—Early Type of Shaft Furnace.

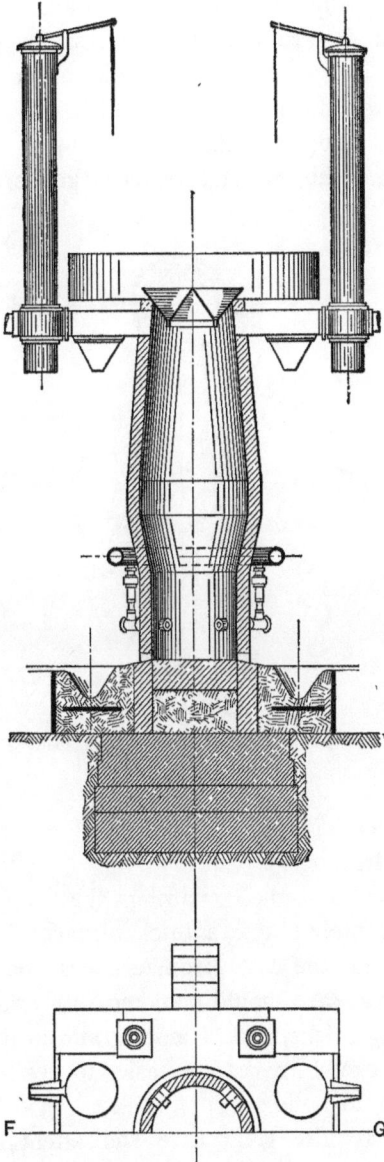

FIG. 98.—Early Type of Shaft Furnace          FIG. 99.—Oker Furnace in the
           Scale, 1 . 150.                              Harz.

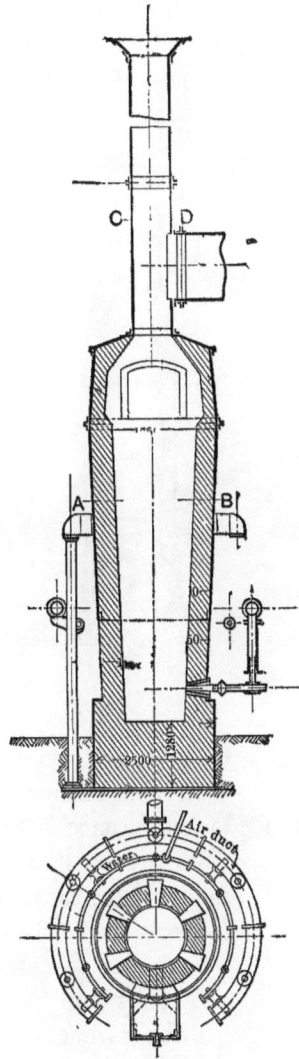

sometimes used, especially for the inside walls. To lessen the danger of interrupting the process by an accident to the water-jacket, the latter should never be constructed of one piece, but in segments which may be independently removed and replaced.

The first requisite for the use of a water-jacket is an

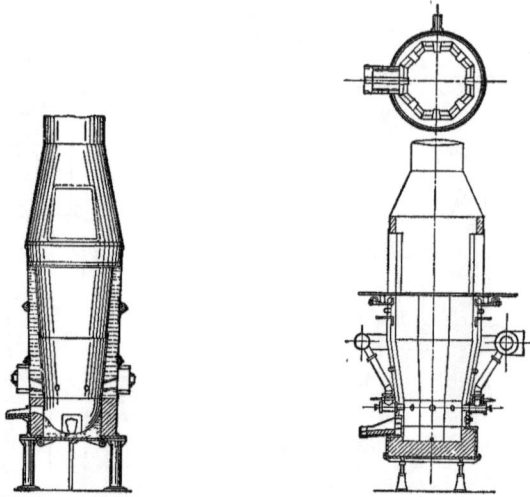

FIG. 100.—Allis-Chalmers.    FIG. 101.—Colorado Iron Works, Denver.
Scale, 1 : 150.

adequate supply of water. If this is available, there is no danger connected with the process.

In construction, operation and repairing, the water-jacket furnace is to be preferred to a brick furnace.

With good management the cost of operation is about the same for each. Great care should be taken with brick furnaces in the choice of bricks, and of mortar, and in the construction. Water-jacket furnaces are easier to design, also easier and quicker to build.

Brick furnaces also require great care and attention when blowing in. If the blast pressure is too high, great difficulty may be encountered, such as burning out the hearth and shaft walls. If the blast pressure is too low, wall-accretions and sows may form. Scarcely any difficulty is experienced in blowing in a water-jacketed furnace.

With similar handling, the relation between the repair costs of brick and water-jacket furnaces is 2:1. Cracks

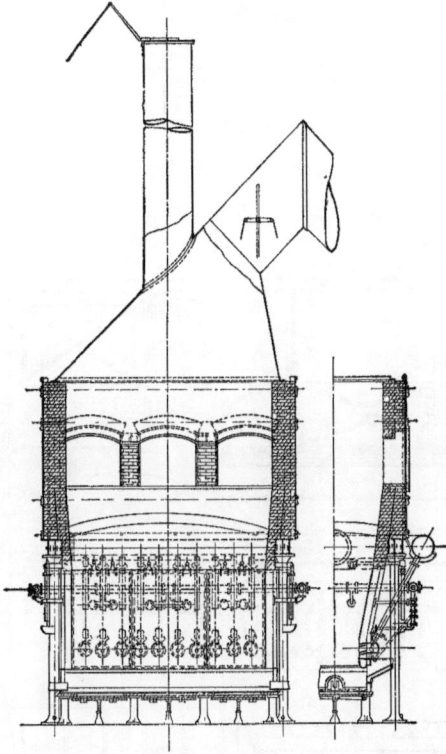

FIG. 102.—American Water-jacket Furnace (20 Tuyeres), Colorado Iron Works. Scale, 1 : 150.

and thin places formed in the walls of brick furnaces tend to grow worse instead of better, for even if they are filled with clay, the matte and slag tend to eat under the clay and widen the crack.

If in order to remove accretions it is necessary to lower the charge, it is often found in brick furnaces that the

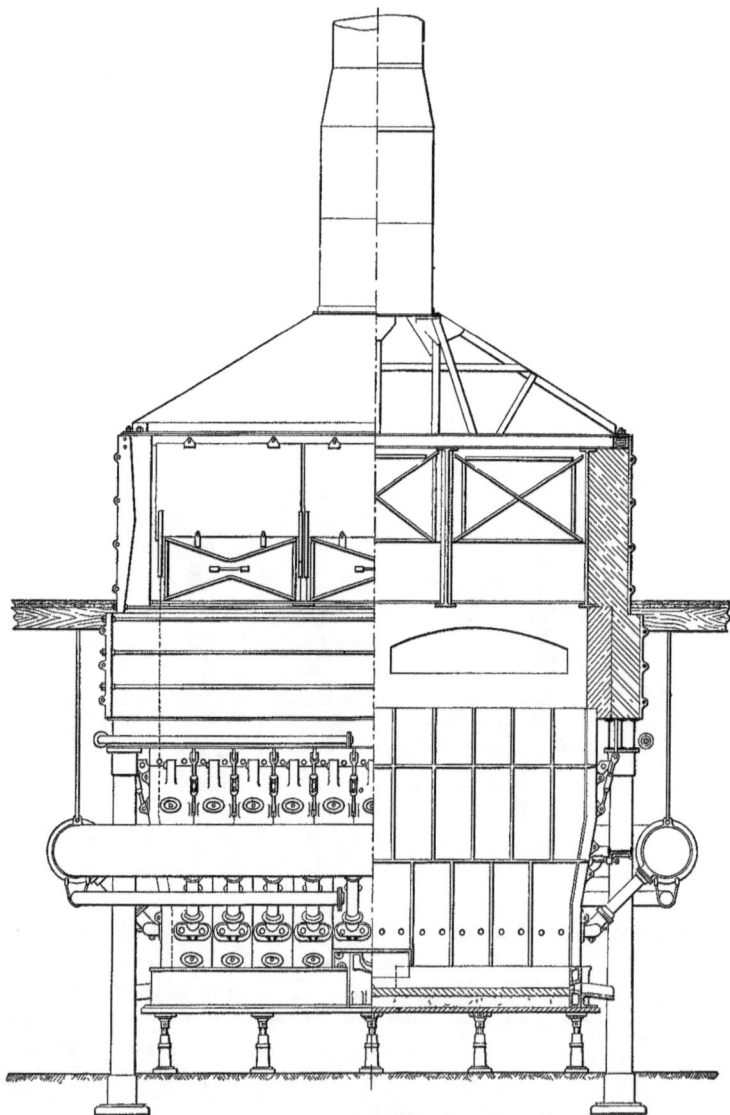

Fig. 103 —Sticht Shaft Furnace, Mount Lyell M & R. Co., Tasmania. Side View and Longitudinal Section.

accretions stick even more firmly to the walls than before, because of the fact that the upper part of the walls of the

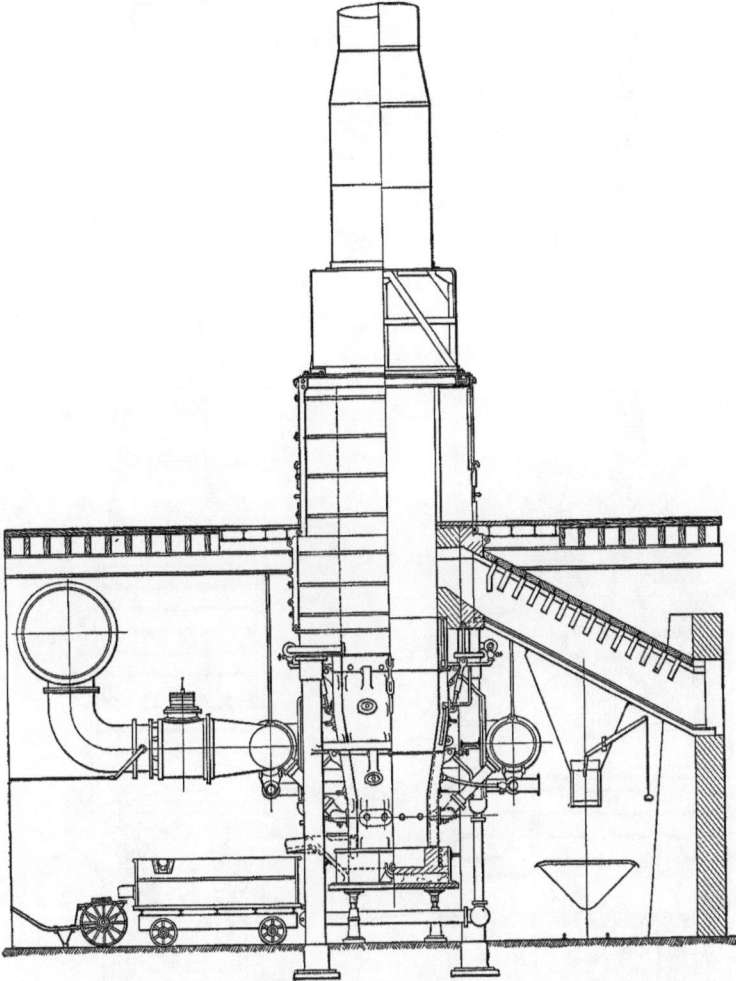

FIG 104 —Sticht Shaft Furnace, Mount Lyell M. & R. Co., Tasmania.   End View and Cross-section.

empty shaft have become very hot.  In the water-jacketed furnace, however, the accretions can be readily removed, for the walls remain cool.

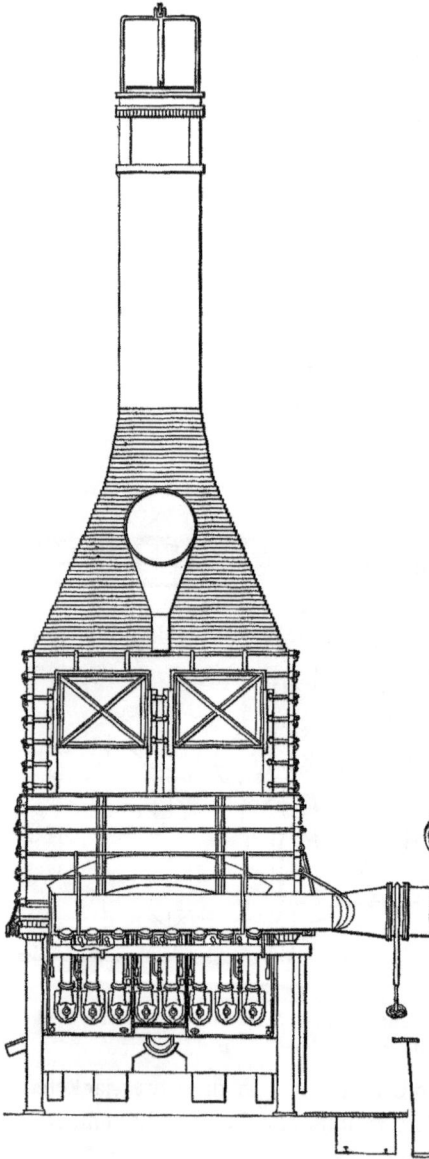

FIG. 105.—Johnson American Water-
jacket Furnace (16 Tuyeres). Side
Elevation.

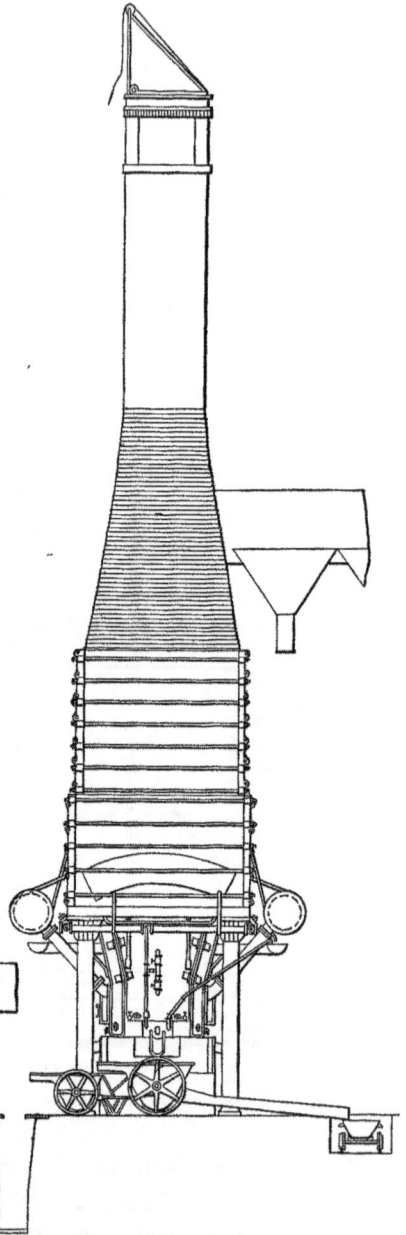

FIG. 106.—Johnson American Water-
jacket Furnace (16 Tuyeres). Side
Elevation.

SLAG AND MATTE SPOUTS: In small furnaces these are made of brasque, in large furnaces of cooled plates formed by fastening wrought-iron tubes, about one inch in diameter, to cast-iron troughs. (Hixon's method of building Lürmann's slag spout.)

FORE-HEARTHS. To lessen undesirable reactions which result in the formation of furnace " sows," copper losses and other difficulties, it is of great importance in copper blast-furnace smelting that the matte be removed as soon as possible from the furnace. The furnaces are therefore almost always built with external crucibles so that the separation of matte and slag may take place in the fore-hearth.

The first fore-hearths were made of brasque in the form of holes in front of the furnace. (See Figs. 97 and 98, pp. 74, 75.)

The modern fore-hearths consist either of portable iron pots or larger boxes (Figs. 104 and 106, pp. 79, 80), or of special reverberatory furnaces for acid slags. The smaller pots are lined with clay and straw, the larger ones are lined a half course of firebrick. If the fore-hearth is very large the lining is omitted entirely. The first slag chills on the sides of the hearth and the heat radiated from the walls so regulates the temperature that the slag remains as a lining. A diameter of 10 ft. is the upper limit for brick-lined hearths.

Water cooling is resorted to in some cases for the whole wall of the hearth as well as for special parts, for example the matte tap, the slag spout, and in the so-called Orford fore-hearth for the partition which separates the matte from the slag.

According to Peters the use of reverberatories as fore-hearths possesses the following advantages:

(a) Economy in the cost of remelting for subsequent treatment of the matte.

(b) Reduction of the necessary temperature required in the blast furnace and thus saving of fuel.

(c) Prevention of overdriving the blast furnace, for if little matte is required at any period of the subsequent operations, the time can be utilized in accumulating a small stock of especially rich or poor matte, casting it into bars of definite size and laying it to one side. A means is, thus provided for regulating the amount and composition of subsequent charges in the reverberatory.

(d) Assistance in the separation of matte and slag. In the unheated fore-hearth, the temperature is continually falling, while in the reverberatory fore-hearth the products of the blast furnace are run into a heated chamber, which favors the separation of matte and slag.

The reverberatories used as fore-hearths do not differ essentially from other copper reverberatories. The following features, however, should be provided:

(a) The matte from the blast furnace should be introduced at one of the ends.

(b) The slag should be withdrawn from the opposite end.

(c) The fireboxes are, therefore, on the sides and should be arranged so as to close easily.

(d) The molten matte should flow as directly as possible from the fore-hearth to the converters.

(e) Sufficient space should be provided to allow 50 tons of matte to be withdrawn and stored in case of an interruption in the subsequent processes.

The matte is withdrawn from the reverberatories either into a sand bed where it cools and is broken into lumps, or else it is run into casting ladles for delivery in the molten state to the next apparatus.

When a reverberatory fore-hearth is not used, the slag is allowed to flow through small fore-hearths (slag pots) in which a small quantity of matte settles. From these slag pots it is withdrawn to be made into building stone if sufficiently acid, or it flows into slag cars for transportation to the dump. If the location and water supply is

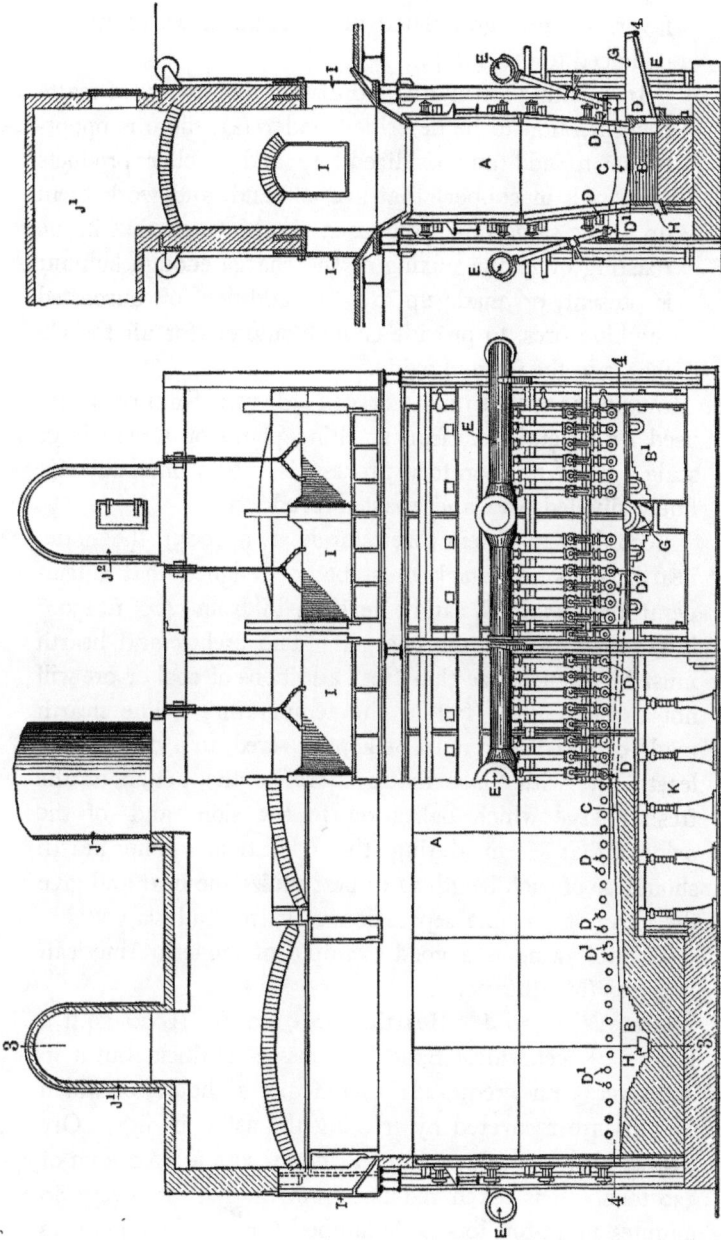

FIG. 107.—Mathewson Furnace made by Building in the Space Between two Furnaces, thus Forming a Furnace with Three Times the Hearth Area of One Original Furnace.

favorable, it is granulated in a stream of water and by it sluiced to the dump.

In smelting for matte, and also in the process of matte concentration to be described under (4), there is opportunity to add any oxidized ore and smelter products (slag rich in copper) that are at hand, and work them up at the same time, but care must be taken that in the roasting or in the mixing of the charge enough sulphur is present, or made up by the addition of unroasted sulphide ores, to provide enough sulphur for all the Cu that is in the form of oxide.

2. REVERBERATORY FURNACES FOR MATTE SMELTING are used at several American smelting plants on a very large scale. The reverberatory process is used principally for finely divided ores and roasted products.

According to Peters (Metallurgie, 2, 9, 1905), the grate, hearth, flues and stack must be so designed that a temperature of 1700° C. can be developed if from 1400° to 1500° is required to fuse the charge. The firebox and hearth must be of such size that fresh additions of coal or ore will not cause a great fall in the temperature. The hearth made of fused sand must be kept covered, to a depth of at least 8 in., with fused matte to resist the action of the fresh charge which bakes on to the sides and of the oxides that form during the operation. The hearth should be of such length as to best utilize the heat and give the cleanest possible separation of matte and slag.

The following is a good example of modern American furnaces (Matthewson):

Grate: 6×16 ft. Hearth: 20×100 ft. (2000 sq.ft.). Flues: 3¼ feet wide. Sand bottom: 3 ft. thick, burnt in one layer. Fuel required: 57 tons per 24 hours, of which 6½ tons are recovered by treating the ashes in jigs. Ore treated averages 275 tons per 24 hours with a maximum of 330 tons. It is taken red hot from the roasters every 80 minutes in 15-ton lots and dumped through four hoppers upon the first 20 feet of the hearth back of the fire bridge. The molten matte, 8 in. deep on the hearth, corresponds to 60-80 tons. When matte is needed in the converter department, it is drawn into ladles, up to 10 tons at a time. Every 3 or 4 hours 30-40 tons of slag are withdrawn,

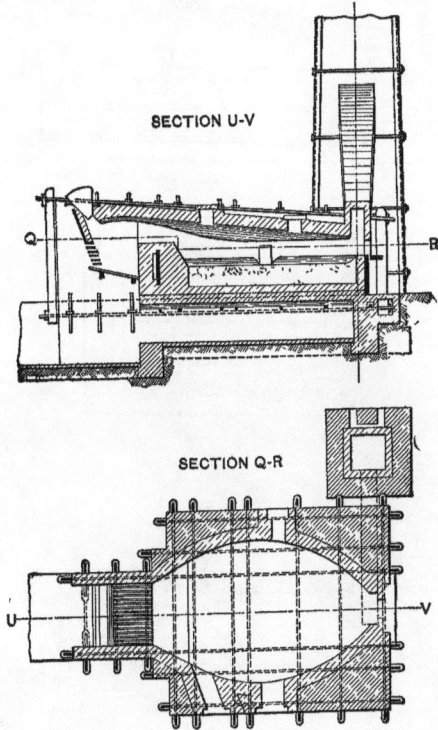

SECTION U-V

SECTION Q-R

FIG. 108.—Mansfeld Furnaces.

requiring 15 minutes for the operation. The slag is granulated and sluiced to the dump.

The process requires little hand labor, as the charge, dumped in the hottest zone and floating upon the matte, distributes itself, and then flows 80 ft. before reaching the taphole. After the large mass has been melted down, the roof and side walls absorb so much heat that at the end of 80 minutes a new charge can also enter the now highly superheated chamber. The process thus becomes continuous as with the blast furnace.

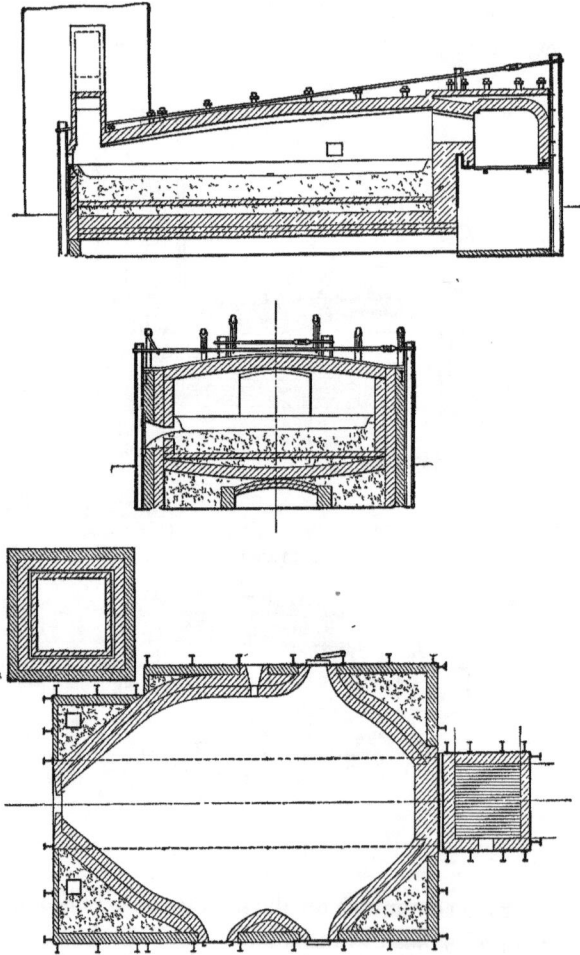

FIG. 109.—American Reverberatory Furnace (Peters). Scale, 1 : 150.

Since the reverberatory furnaces for the different processes in copper smelting (matte smelting, black copper smelting and copper refining) are of similar design, although

of different dimensions, they are shown together on pages
85–87 (Figs. 108–110).

1. **Oxidizing Roasting** and

2. **Smelting for Matte** are often carried on in one operation.
This is exemplified in the old process, kernel roasting, and in
pyritic smelting, a process brought to high efficiency during
the past ten years in Australia and in the United States.

**Kernel Roasting:** This is a prolonged heap roasting, the purpose being to concentrate the copper at the centre of the slowly
roasted lump of ore. By this method of roasting there takes

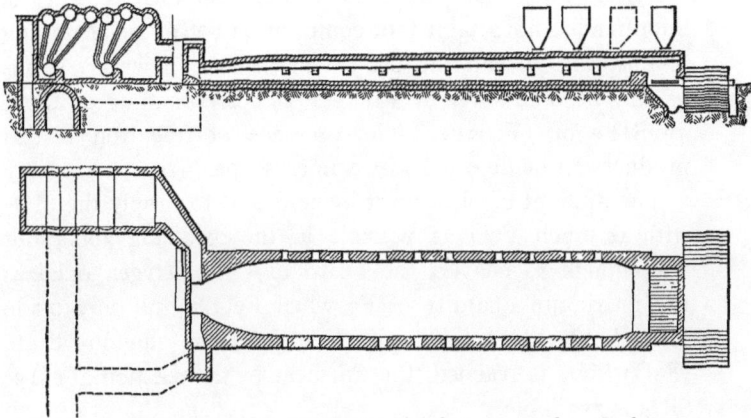

FIG. 110.—Mathewson Furnace, Washoe Smelter, Anaconda    Scale, 1 : 500

place at the contact of the oxidized surface and the unchanged
sulphide a reaction similar to that which occurs in matte
smelting. The copper is taken up by the unaltered sulphide, which results in a constant enrichment of the centre
in copper, leaving a crust composed of iron oxide. At the
temperature maintained in this process, part of the copper
is unavoidably converted into $CuSO_4$, and remains as such
in the crust. The kernel can be enriched to about five times
the copper content of the ore; the crust usually contains about
3% Cu as $CuSO_4$ and 1% as CuO.

**Pyritic Smelting** (Metallurgie, **3**, 1906). True pyritic smelting consists of an oxidizing smelt without using any fuel

except that contained in the ore itself (Fe and S). It is used for pyrite carrying copper and precious metals, but with not too much lead ($<10\%$) or zinc ($<10\%$ of the slag). For ores low in gold there should be 0.3 to 0.5% Cu; for ores rich in gold (13 to 30 oz. per ton) there should be 1 to 3% Cu in order to prevent a high loss in gold.

The S and especially Fe are the sources of heat. One part by weight of O in uniting with Fe (to FeO) evolves four times the amount of heat as when uniting with S (to $SO_2$). To this is added the heat of neutralization of FeO and $SiO_2$.

Pyrite ($FeS_2$), as it descends through the furnace, loses S, but often has not attained the composition FeS at the beginning of fusion. On entering the oxidation zone, however, it has usually become lower in S than FeS, corresponding to about 4FeS:Fe or 3FeS:Fe. The presence of free iron in this product can be detected with a microscope.

The amount of blast must be sufficient to supply the $SiO_2$ with as much FeO as possible in the oxidizing zone, but not enough to convert the Fe to $Fe_2O_3$. Oxygen oxidizes the maximum quantity of Fe when FeO is the only oxide formed. If the ratio required to form the singulo-silicate $(FeO)_2SiO_2$ is reached, the full heat of the reaction $2FeO + SiO_2$ is available.

If there is a lack of $SiO_2$, part of the iron will be converted into the higher oxides ($Fe_3O_4$ or $Fe_2O_3$). These oxides do not themselves form silicates, but absorb heat on being dissolved in the slag; the slag becomes cooler and more viscous.

Hot blast, contrary to many recommendations, should be avoided in most cases. It favors the formation of acid slag and then the energy of saturation of the $SiO_2$ is not fully utilized. Hot blast is of advantage only for ores which are low in pyritic material and require considerable additional fuel.

The concentration ratio (ore: matte) becomes greatest when the slag most nearly approaches a singulo-silicate.

Ores of such composition that pure pyritic smelting is possible are rare, but there are many plants that have reduced the fuel consumption to 1–5% of the total charge, whereas it was 15–20% when using roasted ore.

3. **Oxidizing Roast of the Matte** and

4. **Matte Concentration by Smelting** are merely repetitions of operations 1 and 2, and the object and apparatus are the same. As in the treatment of ores, it has been found possible to unite processes 3 and 4 successfully by an oxidizing smelt. For this purpose either reverberatory furnaces or converters are used.

**Reverberatory Smelting.**

This has been already sufficiently described under 2.

**Converting Copper Matte** is not only carried to the point where the matte is concentrated to $Cu_2S$, but some metallic Cu may be formed. It is merely a combination of the roasting and smelting processes of matte concentration, carried out by blowing compressed air through the molten matte. The matte is thereby desulphurized nearly to the composition $Cu_2S$, the iron is oxidized to FeO and slagged by the silicious lining of the converter. After the matte has been converted to $Cu_2S$ the process is carried out in different ways and this, together with the apparatus, will be described under " (B) Extraction of copper " (page 95).

The concentration of the matte aimed at in these enrichment processes lies between 72 to 78% Cu. Above this point the reactions soon to be described under copper extraction take place. This is due to the fact that when the quantity of FeS becomes small, copper separates out (copper bottoms), together with silver if it is present in large amount.

A third roasting and smelting is to-day seldom performed, although it was earlier the general practice, especially in Wales.

Regarding the nature of copper matte, or rather of the different states of concentration, there was until recently considerable obscurity. Recently several researches have been made on the subject, of which, however, only one, that of P. Röntgen, Aachen, can lay claim to great accuracy. According to this work, there exist several compounds (at least three) between $Cu_2S$ and FeS, one with about 58% Cu corresponding to $(Cu_2S)_3.(FeS)_2$, one with about 50%, corresponding to $Cu_2S.FeS$, and one with about 33% Cu, corresponding to $(Cu_2S)_2.(FeS)_5$. Furthermore there is evidence of a eutectic between $Cu_2S.FeS$ and $(Cu_2S)_2.(FeS)_5$, corresponding approximately to the composition $Cu_2S.2FeS$:

$$Cu_2S.FeS + (Cu_2S)_2.(FeS)_5 = 3[Cu_2S + 2FeS].$$

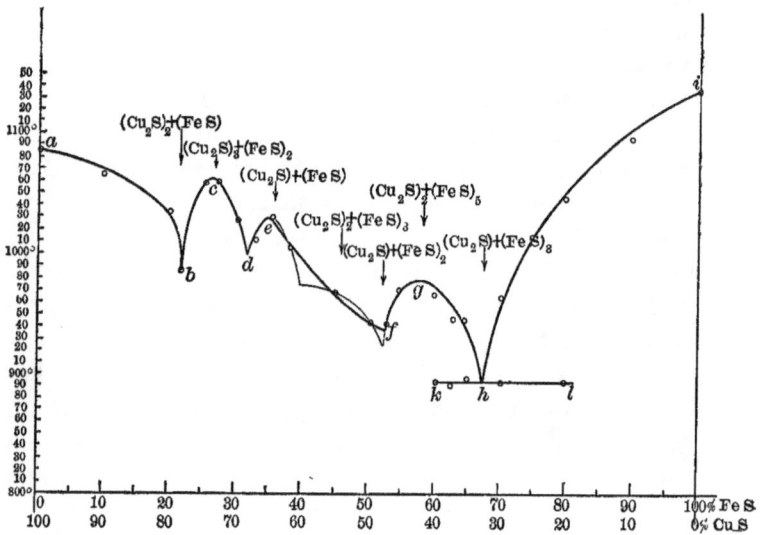

FIG. III.

It is interesting to note that if we imagine the first compound $(Cu_2S)_3.(FeS)_2$ to be more highly sulphurized still, we come to bournite ore $(Cu_2S)_3.FeS.FeS_2$, while the just-mentioned eutectic corresponds to chalcopyrite $Cu_2S.FeS.FeS_2$. (See Fig. III.)

**Concentration of Copper in Aqueous Solution by Leaching,** can be substituted advantageously for the foregoing smelting processes, when the roasted products contain copper compounds (sulphate, carbonate, and oxide) that are readily soluble in water and acids, but not when the relatively insoluble copper

sulphides are united with other sulphides. Pure copper sulphides may be converted into soluble compounds by treatment
with different agents (see below, especially ferric and cupric
salts). Ores and smelter products, on the other hand, in which
the CuO and $Cu_2O$ are combined with other oxides ($Fe_2O_3$,
$SiO_2$, etc.) or in which the $Cu_2S$ is combined with other sulphides
(FeS, $FeS_2$, etc., in chalcopyrite and bournite) are only slightly
attacked by the customary solvents. Double sulphides may be
converted by roasting at a low temperature (450–500° C.) into
insoluble $Fe_2O_3$, soluble $Cu_2O$, CuO, $CuSO_4$ and unoxidized,
but free, $Cu_2S$. Moreover, according to a recent investigation
by O. Frölich, the double sulphide may be decomposed by
heating to about 200° into the simple free sulphides without
forming any oxides:

$$Cu_2S.FeS = Cu_2S + FeS$$

(cf. Metallurgie, 1908, **5**, 206).

The following solvents are used:

**Water for Sulphates.** Since in weathering and roasting some
basic sulphates are formed, the water is usually acidulated
if it is not directly obtainable as acid mine water.

**Hydrochloric Acid** is used ordinarily in connection with other
salts, e.g. (NaCl), to increase the solubility of cuprous and
silver chlorides. At the Stadtberge works in Lower Marsberg,
150 tons of acid schist, carrying 2% Cu as carbonate and as
copper glance, were formerly leached daily as follows: The
ore, crushed to 6–15 mm., was placed in shallow trenches
which had been made water-tight with clay and lined with
boards. Here it was treated for 8–10 days with a solution
of acid and NaCl, whereby copper amounting to 0.5% of
the weight of the ore was dissolved. The residue was then
allowed to stand in the air for from 8–10 weeks in well ventilated heaps and again leached with salt solution (liquors
containing $FeCl_2$ waste from the precipitation vats). This
dissolved 1% Cu, based on the weight of the ore. After the
second leaching the ore was transferred to the dumps, where

it was again subjected to the action of the atmosphere. Here almost all of the remaining copper was leached out by the action of the rain, the solution being removed in drainage trenches and conducted to the precipitation apparatus.

The reactions in the three stages of the leaching are as follows: The acid dissolves the free carbonate. The ore soaked with free HCl, $FeCl_2$ and $FeCl_3$, as it remains in the heaps and later on the dumps, undergoes the following changes:

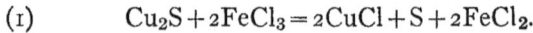

(1) $$Cu_2S + 2FeCl_3 = 2CuCl + S + 2FeCl_2.$$

(2) $$2FeCl_2 + 2HCl + O = H_2O + 2FeCl_3.$$

The $FeCl_3$ thus formed reacts again as under (1). The CuCl may also be converted to the higher chloride:

(3) $$2CuCl + 2HCl + O = H_2O + 2CuCl_2.$$

The latter now works in a similar way to $FeCl_3$.

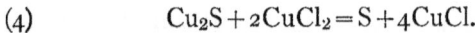

(4) $$Cu_2S + 2CuCl_2 = S + 4CuCl.$$

**Sulphuric Acid.** The most important cases where sulphuric acid is utilized have already been described under Silver.

**Ferric and Cupric Salts.** The action of ferric and cupric chlorides has already been illustrated under hydrochloric acid. Ferric sulphate acts in a similar way, but cupric sulphate does not, because cuprous sulphate is so unstable that it is not formed under the conditions that prevail in leaching.

In the Rio Tinto district (Fig. 112), where pyrite occurs with free $Cu_2S$, the ore is arranged in long, terraced heaps 28 ft. high, placed side by side and provided with air passages. The heaps are sprayed from time to time with acid mine water, to leach out the $CuSO_4$ and to assist in the sulphatizing action. The ferrous sulphate and sulphuric acid contained in the mine water and in the heaps apparently come from a part of the pyrite:

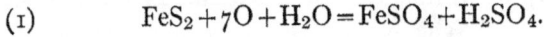

(1)          $FeS_2 + 7O + H_2O = FeSO_4 + H_2SO_4.$

If a considerable quantity is present, it may be oxidized to ferric sulphate:

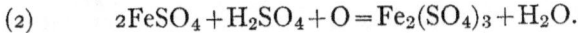

(2)          $2FeSO_4 + H_2SO_4 + O = Fe_2(SO_4)_3 + H_2O.$

FIG. 112.—Rio Tinto Leaching Plant.

The $Fe_2(SO_4)_3$ then reacts with the $Cu_2S$ to some extent as follows:

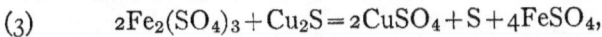

(3)          $2Fe_2(SO_4)_3 + Cu_2S = 2CuSO_4 + S + 4FeSO_4,$

or, together with the atmospheric oxygen and moisture,

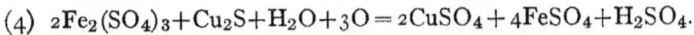

(4) $2Fe_2(SO_4)_3 + Cu_2S + H_2O + 3O = 2CuSO_4 + 4FeSO_4 + H_2SO_4.$

In any case so much $FeSO_4$ is formed, as compared

to the amount of $H_2SO_4$ present, that under a very active oxidation much basic sulphate results and a considerable quantity of hydrated ferric oxide forms on the surface of the ore. · The removal of the copper takes place so rapidly that much $FeS_2$ remains unaltered and the residue from leaching, still containing 49 to 50% S, may be used in sulphuric acid manufacture.

Ferric chloride leaching was at one time introduced into the Rio Tinto district, but the practice has been abandoned (Dötsch Process).

**Ferrous Salts,** especially $FeCl_2$, have been recommended by Hunt and Douglas in leaching ores which contain copper as oxide or carbonate:

$$3CuO + 2FeCl_2 = Fe_2O_3 + CuCl_2 + 2CuCl.$$

As an argument in favor of the process it was pointed out that CuCl solution requires less iron for precipitating the copper than $CuCl_2$ or $CuSO_4$ solutions.

Ores which cannot be leached directly are prepared under some circumstances by subjecting them to a chloridizing or a sulphatizing roast. In both cases an oxidizing roast is carried out; in the first case chlorides ($MgCl_2$, NaCl, etc.) are added and chlorine is made available by the aid of $SO_2$ and O, and in the second case the temperature is kept as low as possible (450 to 500° C).

A PURIFICATION OF THE LEACHED LIQUOR is necessary if the plant also sells copper as blue vitriol. The chief impurity is $FeSO_4$, together with small quantities of Sb, As, etc. All these are substances easily precipitated if copper oxide (roasted matte) is introduced while air is being blown through the hot solution for the purpose of oxidizing $FeSO_4$ to $Fe_2(SO_4)_3$.

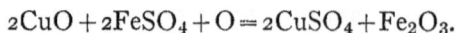

$$2CuO + 2FeSO_4 + O = 2CuSO_4 + Fe_2O_3.$$

This process was brought to a high state of efficiency by Ottokar Hofmann in Kansas City (Metallurgie, 1907, **4**, 582).

# (B) Extraction of Copper from Intermediary Products

In the concentration processes described under $A$, the copper remained either in the form of sulphide low in iron ($Cu_2S$) often carrying precious metals, or in the form of an aqueous solution. The concentrated matte is commonly converted into copper by smelting processes, among which the so-called *Reaction Process* is most commonly employed, although the *Roast-reduction Process* has found favor if a matte containing silver is to be desilverized by leaching between the roasting and smelting operations. If the copper has been concentrated in the wet way, obviously only precipitation methods come into consideration.

**Reaction Smelting** is carried out either in reverberatories or in converters, and causes the separation of copper according to one of the following reactions:

With the sulphides of base metals, the oxidation always takes place so that the metal oxide and $SO_2$ are formed:

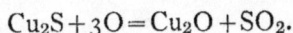

$$Cu_2S + 3O = Cu_2O + SO_2.$$

The higher oxide is not formed in the presence of reducing agents ($Cu_2S$), but the latter, on the contrary, reduces the $Cu_2O$:

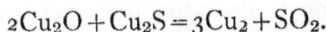

$$2Cu_2O + Cu_2S = 3Cu_2 + SO_2.$$

The reverberatory furnaces used for this process are not so large as the newer smelting furnaces, but like the latter they have a sand bottom and are provided with tuyeres beside the firebox for blowing air upon the metal bath (cf. page 85-87).

In 1880 Manhès in France and one of his assistants at the Parrot Copper Company of Butte, Montana, succeeded in blowing copper matte directly to metallic copper in a converter similar to that used in the Bessemer process. The success was due to a change in the method of admitting the air. The previous failures had been due to using a converter of exactly the same pattern as that used in steel manufacture without taking into account the different conditions governing the working up of the raw materials used in copper smelting. In the copper Bessemer process a matte carrying 40 to 55% Cu is commonly used. In comparison with the iron Bessemer process (see Iron) the following differences may be noted.

| Copper Matte | Iron |
|---|---|
| 45 to 60% oxidizable substances of low calorific power. | 3 to 6% impurities of high calorific power. |
| Formation of large amounts of strongly basic FeO which slags with the converter lining. | Quantity of slag small. |
| Converter lining soon destroyed (9 charges). | Converter lining lasts a long time (200–250 charges). |
| Three different materials: slag, constantly increasing; matte, constantly decreasing, metal, in the second stage constantly increasing. | Two materials: small amount of slag. Large amount of metal. |
| Copper: good conductor of heat Specific heat 0.155 (liquid, see Glaser, Metallurgie, 1904, **1**, 126) | Iron: poorer conductor. Specific heat 0.1665 (liquid, see Oberhoffer Metallurgie, 1907, **4**, 495). |

Only when Manhès arranged the tuyeres so that the air no longer passed through the separated metallic copper was the blowing successful.

MANHÈS' CONVERTER. This consists of a cylinder on a horizontal axis; the tuyeres are arranged in a row, in the cylinder mantle, being placed parallel to the axis, so that by rotating the drum the tuyeres can be made to enter the matte at different depths. This cylindrical type has been successful at both the Anaconda and Copper Queen (Bisbee) plants. Fig. 113 shows a modern arrangement of such a plant.

STALMANN constructed a converter of rectangular cross-section. These converters have been introduced by Sticht at

FIG. 113.—Horizontal Converters at the Anaconda Plant.

FIG. 114 —Converter Department, Mount Lyell M. & R. Co., Tasmania.

the large plant of the Mount Lyell Mining and Railway Co., Australia. Fig. 114 shows the converter department of these

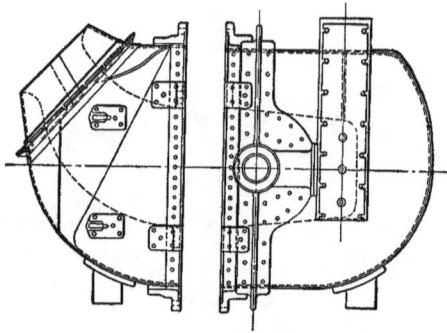

FIG. 116.

FIG. 115.

Stalmann Type of Converter at Mount Lyell. Designed by R. Sticht.

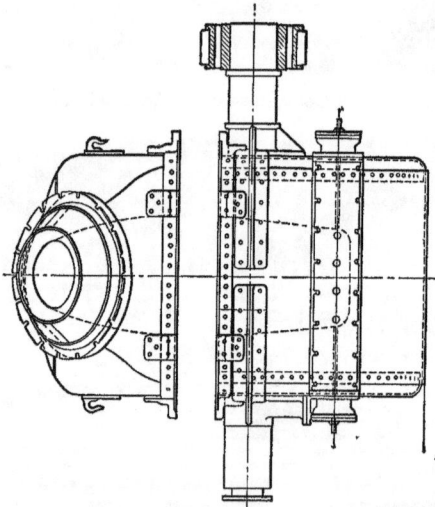

works. Figs. 115, 116 and 117 also show sketches of the converters as made by Mr. Sticht.

Upright converters, resembling the Bessemer type, have been successfully used at the Parrot and Anaconda plants.

The tuyeres are placed in a semicircle above the bottom on one side of the converter (Figs. 118–121, p. 100).

The converter lining, consisting of 17–28% clay and 83–72% quartz (gold bearing), is tamped into a thickness of about 18 in.

The blowing takes place in two stages: (1) Oxidation of the FeS and slagging of FeO, usually requiring 30–50 minutes, and (2) blowing to separate the copper, requiring 30–50 minutes more.

**The Roast-Reduction Process:**

This consists of the following operations:

1. OXIDIZING ROAST (dead roast) for the purpose of removing

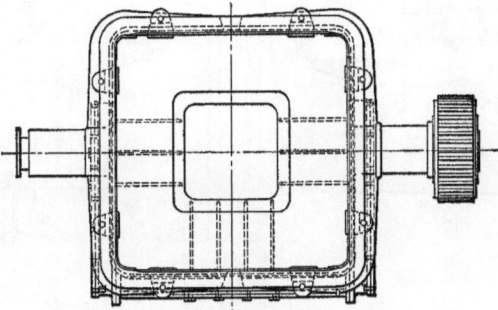

FIG. 117.
Stalmann Type of Converter at Mount Lvell.   Designed by R. Sticht.

as much of the S as possible and converting the Cu into CuO. (Apparatus: see A, p. 61 to 71.)

2. REDUCING SMELT in a shaft furnace or a reverberatory, with other rich copper-bearing materials if desired. This is very successfully used at the Mansfeld works in conjunction with the Ziervogel process. Reverberatory furnaces are used at this plant.

**Precipitation of Copper** from solutions of the chloride and sulphate is almost always accomplished by iron. In the Stadtberge Works (page 91) wooden vats are used in which the iron in the form of thin turnings rests upon a wooden frame and the solution is agitated by wooden paddles. At Rio Tinto the copper is precipitated in trenches in which cast iron bars have been placed. (Fig. 122).

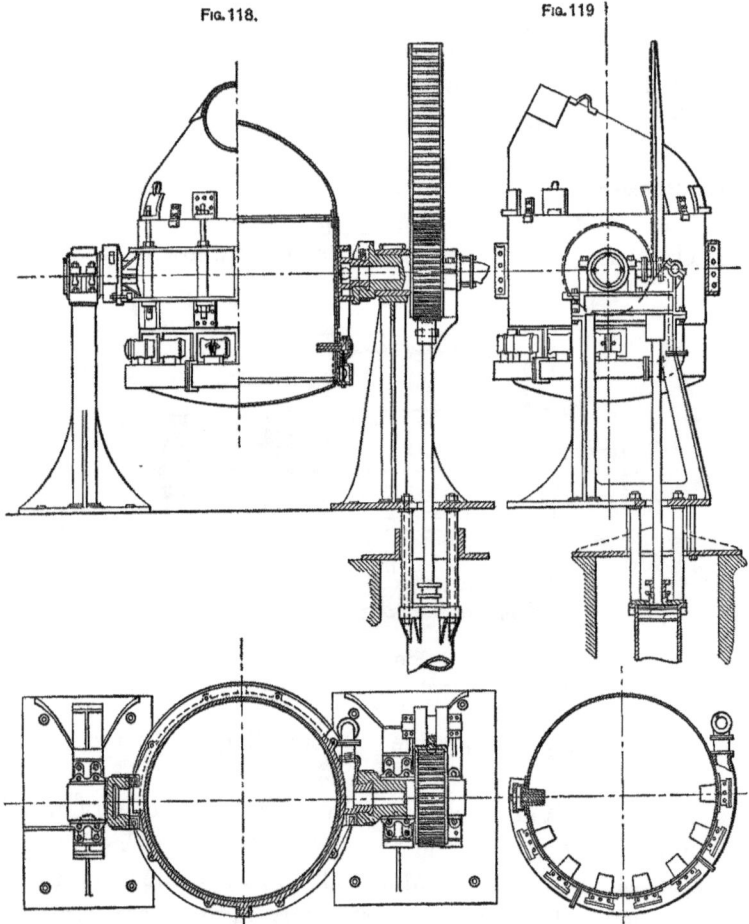

FIG. 118.

FIG. 119

FIG. 120.

FIG. 121.

Converters at the Anaconda Plant.

The precipitated copper is treated in jigs to remove large pieces of iron and in washing drums to remove fine particles of iron and iron oxide. It is then filtered and dried.

If the solution carries silver, part of the copper (one-third to three-eighths) is first precipitated, carrying with it the silver, then the remaining copper is precipitated, silver

FIG. 122.—Rio Tinto Precipitating Trenches.

free, in a second vat. The two lots are treated differently (see below).

# (C) Copper Refining

Pure copper, or refined copper, is made to-day almost entirely from crude copper. The attempts to produce pure copper directly from the ore have thus far proved a failure, and the same is true of most attempts to produce pure copper from matte. It is only recently that the difficulties hitherto met with in the latter process have been overcome.

Crude copper may contain the following impurities: S, As, Sb, Zn, Pb, Fe, Ni, Co, Ag, Au. If precious metals are not present the following process may be used:

**Furnace Refining** consists of an oxidizing followed by a reducing fusion.

1. OXIDIZING FUSION. A reverberatory furnace is usually used for this purpose, the object being to eliminate the oxidizable impurities from the copper. The S, As, Sb, Zn, and Pb are partly volatilized as oxides and partly slagged; any Fe and Ni are also slagged. The S is present in impure copper largely as $Cu_2S$, and is removed only after the greater part of the other impurities has been eliminated. It is oxidized principally through the agency of $Cu_2O$ after a certain quantity of the latter has been dissolved in the copper. When this point is reached, the evolution of $SO_2$ takes place rapidly (boiling); when the reaction has nearly ceased, a wooden pole is plunged into the metal.

The gases escaping from the submerged end of the pole agitate the metal, thus allowing the $SO_2$ to escape more easily and at the same time the spurting copper is further oxidized (dense poling). The copper resulting from this treatment contains much $Cu_2O$ (set copper). The resulting slag runs high in copper and goes back to the smelting furnace. Set copper is very brittle and unfit for the ordinary purposes for which copper is utilized. It is further refined as follows:

2. REDUCING FUSION. The molten copper is heated with a reducing flame and kept covered with charcoal; a wooden pole (green wood) is introduced and poling continued until the dissolved $Cu_2O$ has been reduced by the hydrocarbons in the pole and the charcoal in the covering. This process is complete when a thin test-bar can be bent through 180° without showing a crack. When broken by repeated bending, the fracture must show a fibrous, salmon-red texture with silky luster.

The refined copper is cast from wrought-iron ladles, which are coated with chalk or clay, into cast iron or copper

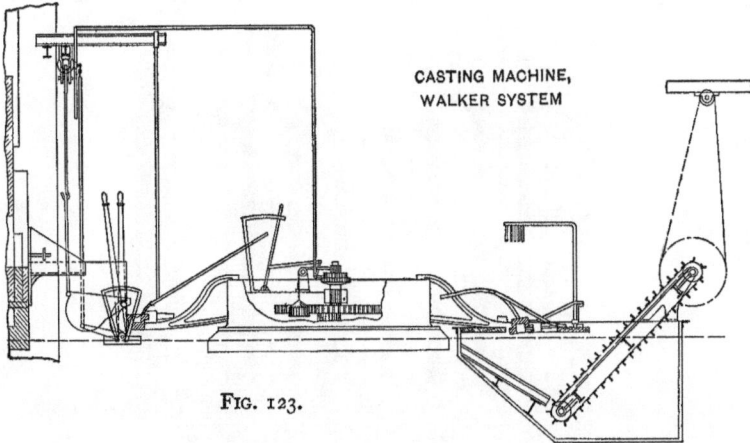

CASTING MACHINE,
WALKER SYSTEM

FIG. 123.

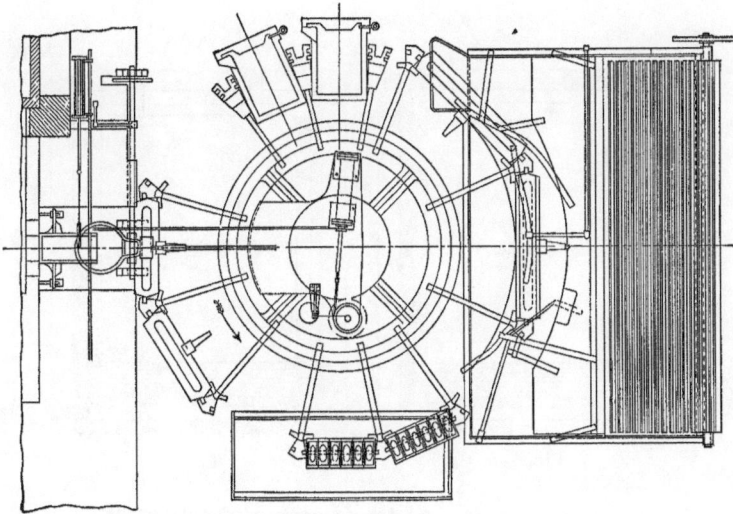

FIG. 124.

molds. In large plants it is run directly from the furnace through troughs into mechanically dumping molds (Figs. 123–124).

**Electrolytic Refining** is almost universally used for copper containing precious metals. It is carried out in ELECTROLYZING VATS made of wood but lined with lead.

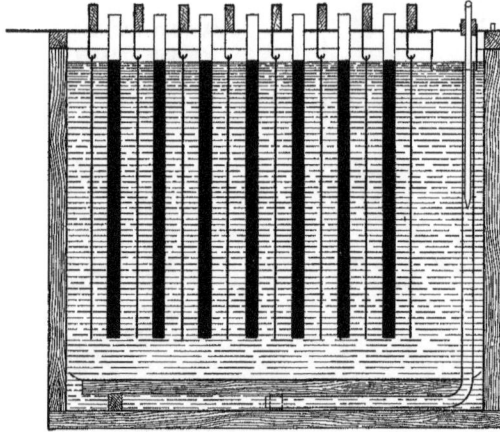

FIG. 125

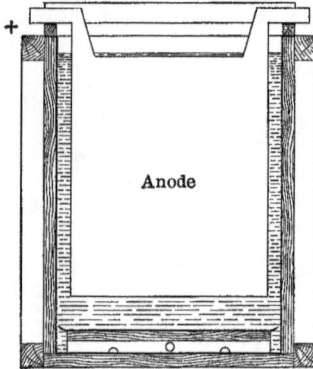 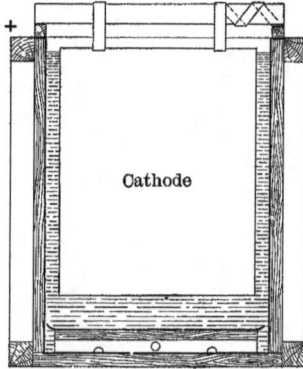

FIG. 126.       FIG. 127.

THE ANODES are cast plates of unrefined copper (Figs. 125, 126, and 127) that contains precious metals.

THE CATHODES are thin sheets of copper made by depositing the metal electrically on lead or greased copper plates from which it can be easily removed.

THE ELECTROLYTE is an acid solution of copper sulphate containing 12–15% $CuSO_4.5H_2O$ and 5–10% $H_2SO_4$.

DURING THE PROCESS, it is important that the constantly decreasing quantity of free acid and the constantly increasing quantity of Cu in the electrolyte does not get outside the above limits. More copper is dissolved from the anode than corresponds to the current density, because of the action of the atmospheric oxygen:

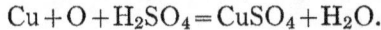

$$Cu + O + H_2SO_4 = CuSO_4 + H_2O.$$

The above reaction takes place under normal conditions only at the surface of the anode. If, however, the process is carried on at a low current density, a part of the copper as it goes into solution from the anode may become incompletely charged, assuming a univalent instead of a bivalent charge; in other words, the solution then contains cuprous ions, but these are precipitated again at the anode:

$$Cu^{\cdot} + Cu^{\cdot} = Cu^{\cdot\cdot} + Cu.$$

It follows, therefore, that the insoluble particles remaining at the surface of the anode, which fall to the bottom as slime, are likely to contain considerable copper. The latter, however, will be easily dissolved if the electrolyte is agitated with air so that the concentration at the electrodes is equalized.

THE TEMPERATURE may be raised to 40° C. (104° F.); there is no advantage in raising the temperature above this point, if the quality of the cathode copper is taken into consideration.

THE CURRENT DENSITY may vary, according to the quality of the anode copper, between 4 and 15 amperes per sq.ft.

THE POTENTIAL should be 0.1–0.3 volt.

THE REACTIONS DURING ELECTROLYSIS are as follows: The anode copper carrying a positive charge goes easily into solution and passes to the cathode, where it loses its charge and is deposited. Of the impurities present in anode copper, As, Sb, Bi and Sn go partly into solution, but are again partly precipitated; Ag, Au, Pt, and $Cu_2S$ are insoluble; Fe, Zn, Ni and Co go into solution, but are not deposited on the cathode. The insoluble substance falls in a finely-divided form to the bottom near the anode, while the soluble substances that are not deposited remain and gradually contaminate the electrolyte, so that it must be drawn off and its valuable constituents recovered.

PRODUCTS OF ELECTROLYTIC REFINING:

Cathode copper (electrolytic copper), 97–99% of the anode copper;

Precious metals,
Bismuth, tellurium, etc., } from the anode mud.
Blue vitriol.

Impure cement copper (arsenical), finally also nickel vitriol. } from the anode mud and the impure electrolyte.

**Electrolytic Treatment of Copper Matte** (Process of Borchers, Franke and Günther) has been carried out since 1907 in an experimental plant at the Mansfeld works. In previous unsuccessful attempts low-grade matte was used, but in this case matte carrying 72% Cu and upward is employed. In its essential features, the method of working is the same as that in the electrolytic refining of metallic copper that carries precious metals. The principal differences in procedure and the products are as follows:

ANODES, cast plates of concentrated matte ($Cu_2S$).

POTENTIAL: higher than in treating metallic copper, viz., 0.75 volt.

REACTIONS DURING ELECTROLYSIS: with the proper current density, the copper is dissolved as when the anode is crude copper, and the sulphur from the $Cu_2S$ remains behind in the free state.

The other constituents of the matte act the same as in metal refining. If the current density is too low, only half of the copper is dissolved, the balance remaining as blue CuS.

PRODUCTS OF THE PROCESS: Besides the products obtained from treating metal, there is in addition,

Sulphur, which is recovered as yellow crystals on drying the anode mud.

Although this process requires more power it possess the following advantages:

ELIMINATION OF SMELTING the concentrated matte to

FIG. 128.

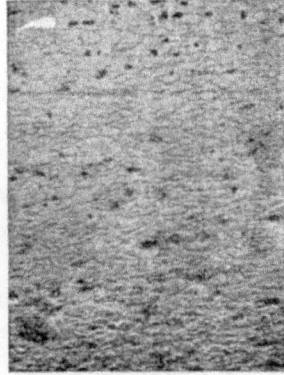

FIG 129.

black copper, and elimination of the damage due to the fumes from such work;

LESSENING OF LOSSES in precious metals, which are greater in smelting for black copper than in any other stage of copper treatment.

RECOVERY OF THE SULPHUR contained in the matte.

## Properties of Refined Copper:

Specific gravity: 8.94.

COLOR: yellowish-red, brilliant luster.

MECHANICAL PROPERTIES: of medium hardness and tensile strength, very ductile.

STRUCTURE of cast and electrolytic copper, granular; of rolled and hammered copper, fibrous (see Figs. 128–129).

MELTING-POINT: 1084° C. (1983° F.).

BOILING-POINT: is said to be in the vicinity of 2100° C. (3815° F.).

THERMAL AND ELECTRICAL CONDUCTIVITY: nearly as great as Ag (0.96).

ALLOYS easily with Mg, Al, Mn, Zn, Cd, Co, Ni, Hg and the precious metals, slightly with Fe, Mo, W, Cr if the metal is pure, more readily in the presence of Si. Copper also alloys easily with its own compounds, especially with $Cu_2O$ and $Cu_2S$, and, when in the molten state, it alloys somewhat with H, CO and $SO_2$.

CHEMICAL BEHAVIOR. In the solid state it is fairly resistant to dry oxygen at the lower temperatures, but above 400° C it is easily oxidized. Crusts of CuO and $Cu_2O$ (copper hammer scale) form on the surface. In a damp atmosphere it easily forms basic salts with O and weak acids (verdigris).

In the molten state copper has a great affinity for S, the the highest of any metal except manganese. Of the acids, only the oxidizing acids, $HNO_3$, hot concentrated $H_2SO_4$, and aqua regia, serve as solvents. With access of air, dilute and weak acids attack copper. Copper does not evolve hydrogen from acids, as its solution tension toward hydrogen is −0.329 volt.

# BISMUTH

## Sources

**Natural Sources:**

FREE, with small quantities of S and As.

MINERALIZED as

OXIDE, $Bi_2O_3$ in bismuth ochre, and as

SULPHIDE, $Bi_2S_3$, in bismuth glance. The sulphide occurs free, in solution or in combination, with sulphides or arsenides of cobalt, nickel, copper, lead and silver.

**Other Sources:**

LITHARGE and other products from the cupelling process.

REFINERY SLAG from silver refining. Residues from the preparation of pure bismuth salts. These may be pure products for which there is no demand or salts that have become contaminated with impurities.

## (A) Concentration Processes

**Mechanical Concentration** consists usually in hand picking.

**Chemical Concentration** is used in working up litharge that contains bismuth. In the section on Silver, it was stated that at the beginning of the cupellation process, when bismuth is present in the bullion, only the lead is oxidized; later, as the bismuth becomes more concentrated in the metal bath, it is also oxidized, forming $Bi_2O_3$, which enters the litharge. The latter seldom reaches a concentration of more than 1 to 2% Bi. It is prepared for the regular Pb-Bi separation by:

1. **Reducing Fusion** to form a Bi-Pb alloy, which is low in Bi at the start.

2. **Oxidizing Fusion,** whereby litharge is formed which is at first low in bismuth but later becomes richer. The different lots of litharge are kept separate and operations 1 and 2 are repeated with the litharge low in Bi. The rich litharge (more than 20% Bi) is treated by leaching and precipitation, usually by

3. **Agitation with Dilute HCl** (15%) accompanied by blowing steam through the solution; this results in the changing of the oxide to chloride.

   APPARATUS: earthenware jars with wooden covers and spigots.

   PRODUCTS: an acid $BiCl_3$ solution containing a small amount of $PbCl_2$ and, in suspension, some undissolved $PbCl_2$ which is removed by

4. **Filtration.**

5. **Stirring** the $BiCl_3$ solution into water and neutralization of most of the free acid with $Ca(OH)_2$ or $Na_2CO_3$. This causes the hydrolysis of the $BiCl_3$:

$$BiCl_3 + H_2O = BiOCl + 2HCl.$$

The BiOCl is precipitated, while the $PbCl_2$, not previously removed by filtration, remains dissolved in the weakly acid solution. The BiOCl is filtered, dried, and is reduced to Bi or is used in fire refining bismuth that carries lead.

# (B) Extraction of Bismuth

**Liquation:** Ores carrying native bismuth were formerly subjected to liquation, in crucibles or retorts, to remove the greater part of the bismuth without melting the gangue. The process can be used only on rich ores and the gangue usually retains 4–5% Bi. The gangue is re-treated by one of the following processes:

**Reduction Process:** applicable to oxidized ores, intermediate products and residues. Among the two latter are bismuth salts, e.g., the oxychloride. Because of the volatility of $Bi_2O_3$

and of Bi itself, and because of the ready reducibility of $Bi_2O_3$, the temperature should be kept low during this process. It is necessary, therefore, to produce a slag of the lowest possible melting point in order that it may be as free from bismuth as possible. The silica content of such a slag should lie between a singulo and bi-silicate, in fact as near to the latter as possible. The bi-silicate slags used in other smelting processes (Cu, Pb), which contain chiefly FeO and CaO as bases, are too infusible for this process. In this case, therefore, $Na_2CO_3$ is used as a flux, besides materials carrying FeO and CaO. The high price of Bi permits the use of this somewhat expensive flux, especially as the fluidity of the slag is such that the $Bi_2O_3$ and the reducing carbon quickly sink through it, thus reducing to a minimum the loss by dusting and volatilization. If basic bismuth salts, such as BiOCl, are to be treated, they should not be added directly to the melted charge because of their volatility. They should rather be stirred into wet $Ca(OH)_2$ or $Na_2CO_3$, dried and calcined for a time in order that the Cl may be completely taken up by the Ca or Na.

APPARATUS: the process may be carried out either in crucibles or in reverberatory furnaces.

**Melting in Crucibles** does not require highly refractory materials, and hence well-baked clay crucibles may be used, although it should be borne in mind that the slag produced in the process will dissolve basic as well as acid oxides. For making these crucibles, which may be done as part of the process, it is important to use as a foundation a clay which has been well baked and subsequently ground. Such material is mixed thoroughly with the least possible amount of fresh clay for a binder. The crucibles are usually formed by hand. For heating, a simple air furnace may be used, but for large plants small continuous kilns are preferred, since they allow a better control of the temperature and are not so hard on the crucibles. The latter is an important point because the use of a graphite crucible, which more easily resists temperature changes, is prohibited. The corrosive action of the

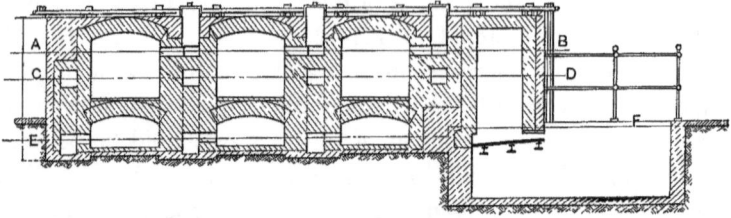

FIG. 130.

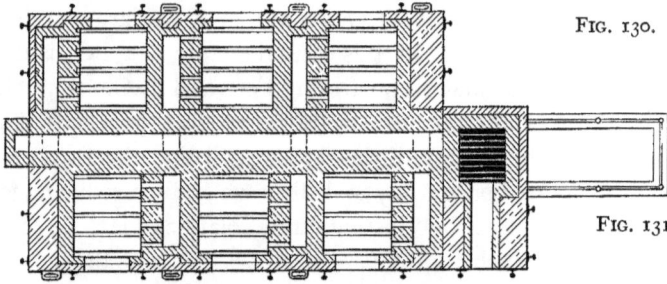

FIG. 131.

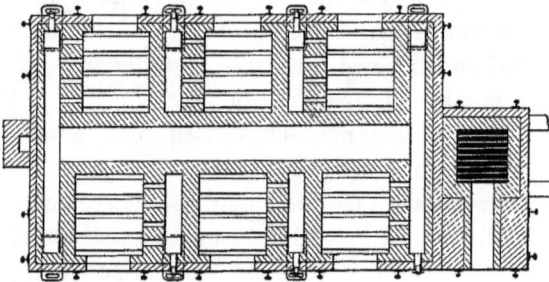

FIG. 132.

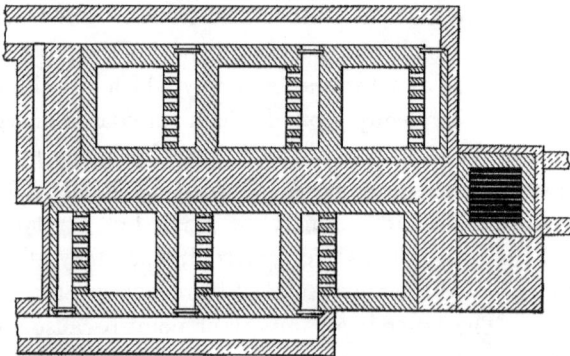

FIG. 133.

Borchers' Continuous Kiln.

slag permits the use of a crucible only once, and the expense of
graphite crucibles would greatly increase the cost of the
process. The common clay crucibles fulfil all requirements
if they are carefully handled. This is difficult in air furnaces, in which the fuel comes in direct contact with the
crucible, where a cold piece of coke touching the hot wall
may cause it to crack. A small continuous kiln designed
by Borchers, with six chambers heated with producer gas,
which is generated in a producer attached to the kiln, completely overcomes the difficulties encountered in the air
furnace and also results in a saving of fuel.

In the continuous kiln shown in Figs. 130 to 133, the hot
gases are conducted from the producer through the main
flue which runs along the center of the upper part of the furnace. From this flue the gases can be conducted by means
of ∩-pipes of sheet iron, fitting into corresponding faucet
pipes, to a branch flue for each chamber. Sliding doors
have not been used here, as they cannot be fitted tightly
enough to prevent the gas from entering chambers which
are cooling. From each of the branch canals a number of
small slits enter the corresponding chamber; between the
slits are openings from an air flue lying beneath. The combustion begins in the chambers above the crucibles and the
hot gases surround the crucible from top to bottom. The
crucibles themselves rest upon supports, between which are
openings that allow the gases to pass into the space below
and thence through connecting flues to the next chamber,
where they warm the filled crucibles. The gases usually
pass through a second chamber before leaving the furnace.
The air is not admitted directly to the combustion chamber
but is first passed through two chambers containing crucibles
filled with hot finished product. In cooling these crucibles
the air becomes pre-heated. From Figs. 130 to 133 it is
seen that six chambers are so arranged that an uninterrupted
continuous process can be carried on. As soon as the contents of one chamber are melted, gas and air are admitted

into the next chamber, which has already been strongly pre-
heated, and the products of combustion are conducted
through an intermediate, freshly-charged chamber. The
chamber in which the air first entered is now closed, the
crucibles are removed, and a new charge is put in place.

**Melting in Reverberatory Furnaces.** In constructing the reverberatory, it should be borne in mind that bismuth in the

FIG. 134.                          FIG. 135.

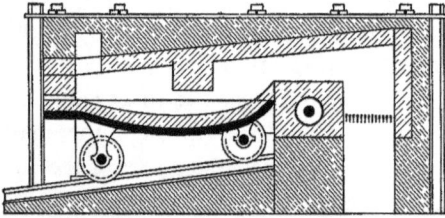 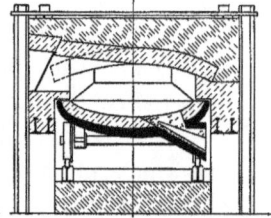

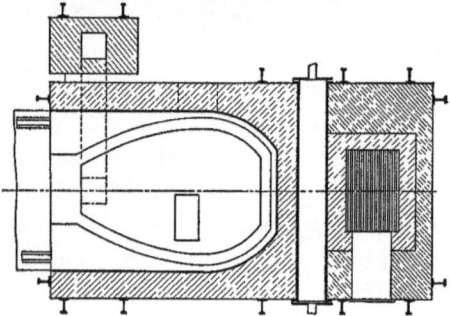 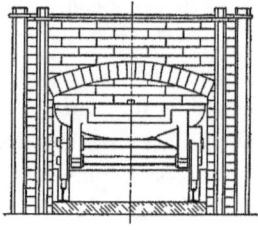

FIG. 136.                          FIG. 137.
Borchers' Reverberatory Furnace.

molten state easily leaks out through the finest cracks or imper-
fections in the walls. Furnaces of earlier construction, having
heavy brickwork supports for the hearth, are being discarded
because they absorb large quantities of bismuth, thus diminishing the output until the furnace is torn down. Even then the
recovery of the metal from the hearth material is tedious and
costly. In plants designed and constructed by Borchers, furnaces have been built with movable iron hearths lined with
half a course of bricks (Figs. 134 to 137). The advantages

of these hearths are a better and quicker extraction of the bismuth and easier access to the hearths for repairs to the lining, without tearing down the side walls. The hearths can be easily changed. Moreover, firebox and flue should be isolated from the hearth walls by an air space or a water-cooled plate, as otherwise the bismuth would flow into the firebox or flue through the two ends that are heated by the fire or flue bridge.

At the beginning of the process in the reverberatory furnace, some slag is first melted and then the fresh charge is introduced. This should if possible sink at once to the bottom, to prevent loss by dusting and volatilization.

**Precipitation of Bismuth** is based upon the decomposition of $Bi_2S_3$ by means of Fe.

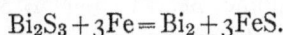

$$Bi_2S_3 + 3Fe = Bi_2 + 3FeS.$$

As the equation shows, this process is applicable to sulphide ores. Apparatus and operations are the same as in the reduction process. If the ores contain substances which may be converted into matte (Cu, Co and Ni) or into speiss (As, Sb) the slag should not be as acid as in the reduction process. The silica content may lie nearer that of a singulo-silicate.

# (C) Bismuth Refining

Crude bismuth usually contains As and Sb, often Pb, and occasionally precious metals. Of these impurities, As and Pb, also Sb if it is present, in small quantities, may be removed by an

**Oxidizing Fusion.** The choice of an oxidizing agent depends upon the nature and amount of the impurities. If much lead is present this must first be removed, as otherwise lead compounds (plumbates) are formed which favor the oxidation of large quantities of bismuth. The lead is best removed by

MELTING WITH BiOCl in iron kettles under a neutral cover of NaCl and KCl, or by

MELTING WITH NaOH with the addition of $NaNO_3$, so that As can be oxidized to $Na_3AsO_4$ in the same apparatus.

In the above operation the NaCl and KCl or the NaOH is first melted, then the metal added, followed by the oxidizing agent, BiOCl or $NaNO_3$, which is stirred in with an iron spatula.

When the refining is completed, an iron hook is lowered in the molten metal and the kettle is cooled. After the mass has solidified, the slag is dissolved in hot water, the kettle is warmed to free the metal from the walls, and this is then lifted out by means of the imbedded hook.

**Melting with Sulphur** and $Na_2CO_3$ or $K_2CO_3$ is used in exceptional cases when the metal is high in Sb. The apparatus and procedure are the same as above.

**Electrolysis** is used when the bismuth carries precious metals.

VESSELS: stone jars.

ANODES: plates or blocks of Bi carrying precious metals.

CATHODES: pure Bi.

ELECTROLYTE: $HNO_3$ and $Bi(NO_3)_3$, or HCl and $BiCl_3$, in aqueous solution.

CURRENT DENSITY: 15–30 amperes per sq. ft.

POTENTIAL: 0.5–1 volt.

## Properties of Refined Bismuth:

SPECIFIC GRAVITY: 9.74–9.8.

COLOR: bright gray, slightly reddish luster.

MECHANICAL PROPERTIES: very brittle, easily pulverized.

STRUCTURE: large-faced, isometric, crystalline grains having a characteristic dendritic structure (Fig. 138).

MELTING POINT: 268° C. (514° F.).

VAPORIZES at red heat.

ELECTRICAL CONDUCTIVITY: .013 compared with Ag.

ALLOYS with most metals. The alloys with Pb, Sn, Zn and Cd have very low melting points, those with Cu and Ni are very hard.

CHEMICAL BEHAVIOR: Bismuth is very resistant toward oxygen at ordinary temperatures, but is easily attacked in the fused state, though not so readily as lead. It is precipitated from its dissolved or fused salts by lead. It is dissolved by

FIG. 138 (×8).

$HNO_3$; by HCl and $H_2SO_4$ only in the presence of oxidizing agents. Hot concentrated $H_2SO_4$ dissolves Bi but the acid is partly reduced to $SO_2$. In all compounds of technical importance Bi is present as trivalent cation.

# LEAD

## Sources

**Natural Sources:**

GALENA (PbS, 86.57% Pb): the associated minerals and gangue are: metallic sulphides such as sphalerite, pyrite, argentite and the sulpho-salts of arsenic and antimony; carbonates, such as cerrusite, smithsonite, limestone and dolomite; further, hematite and also sandstone. The silver content of galena varies between 0.01 and 1%; in the Rhenish provinces 0.01 to .015%; in the Hartz 0.05–0.1%.

CERRUSITE ($PbCO_3$, 77.52% Pb) occurs as an alteration product of galena. It is found in the upper parts of galena deposits and is associated with the same gangue.

**Other Sources:**

LITHARGE: PbO from cupellation.

HEARTH MATERIAL (clay, marl, cement and less often bone ash) saturated with litharge from cupellation.

DROSS, the first oxidation product from lead refining or cupellation, a mixture of lead oxide and lead antimoniate.

LEAD MATTE, an intermediate product from smelting lead ores after the roast-reduction and precipitation processes; contains lead sulphide, iron sulphide, copper sulphide, etc.

Slags which contain an appreciable amount of lead are also returned to the smelter.

ALLOYS of lead with zinc, copper, bismuth, gold and silver.

## (A) Concentration Processes

**Mechanical, Especially Wet Concentration,** is used for ores carrying galena. Because of the high specific gravity of galena, it is easy to concentrate such ores up to a lead content of 70–80%.

# (B)  Extraction of Lead

**Roast-reaction Process.**  By this method of working, sulphide lead ores are said to be treated in such a way that a part of the sulphide is converted by oxidation into oxide and sulphate with which the unchanged sulphide reacts to form $SO_2$ and Pb.  The chemical reactions taking place in the roasting and the reaction-smelting are explained by the following chemical equations:

**Roasting:**

$$PbS + 3O = PbO + SO_2 \qquad PbO + SO_2 + O = PbSO_4.$$

If lead carbonate is also present in the ores, it is decomposed as follows:

$$PbCO_3 = PbO + CO_2.$$

**Reaction Smelting:**

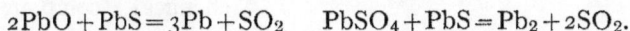

$$2PbO + PbS = 3Pb + SO_2 \qquad PbSO_4 + PbS = Pb_2 + 2SO_2.$$

If the unchanged sulphide and the oxidation products are not present in the above proportions, an excess of oxide or sulphide, as formed by the first equation, will remain unchanged.  The excess of sulphate, however, works as follows:

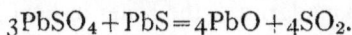

$$2PbSO_4 + PbS = Pb + 2PbO + 3SO_2.$$

$$3PbSO_4 + PbS = 4PbO + 4SO_2.$$

If carbonate is still present during the reaction-smelting it will be changed as follows:

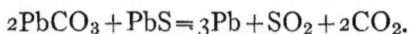

$$2PbCO_3 + PbS = 3Pb + SO_2 + 2CO_2.$$

Schenck and Rassbach have determined experimentally

the conditions of equilibrium between Pb, S and O.  Taking into consideration the $SO_2$ pressures, the conditions given in Fig. 139 were obtained for the formation of $PbSO_4$, PbO and Pb.

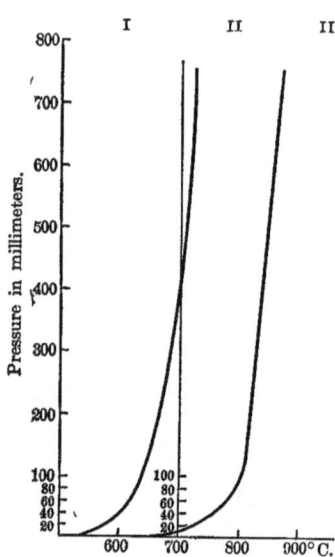

FIG. 139.

Field I is the formation zone for $PbSO_4$; here the principal reaction is $Pb_2 + 2SO_2 = PbSO_4 + PbS$; even PbO is converted into $PbSO_4$ in this zone:

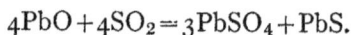

$$4PbO + 4SO_2 = 3PbSO_4 + PbS.$$

Field II is the formation zone for PbO, although the formation of $PbSO_4$ is not yet impossible.  Between PbO and $PbSO_4$, solutions and chemical compounds are formed.  The main reactions here are:

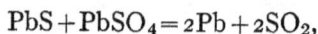

$$PbS + PbSO_4 = 2Pb + 2SO_2,$$

$$3Pb + SO_2 = PbS + 2PbO,$$

and then the following equation,

$$PbS + 3PbSO_4 = 4PbO + 4SO_2.$$

The exact boundaries of the zones in field II, in which on the one hand PbO and $PbSO_4$ form, and on the other PbO alone, have not yet been determined.

Field III is the zone in which Pb alone forms from $PbSO_4$, PbS and PbO:

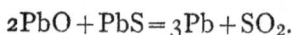

$$PbSO_4 + PbS = 2Pb + 2SO_2.$$

$$2PbO + PbS = 3Pb + SO_2.$$

The roast-reaction process is applicable only to ores rich in lead with not over 4% $SiO_2$. Various modifications in the methods used have sprung up because of local conditions, especially with regard to wages and cost of fuel. These processes are:

CARINTHIAN PROCESS. This is carried out in small reverberatories and therefore with small charges. A low temperature is maintained and the roast and reaction processes are separate. The advantages of this method are: low lead loss by volatilization; little residue; pure lead. Its disadvantages: high fuel consumption and high cost for labor.

ENGLISH PROCESS. Here large reverberatories, large charges, and high temperatures are used from the start. The fuel consumption and cost of labor are low compared with the Carinthian method, but there is a high loss by volatilization, requiring expensive condensation devices. There also remains a large amount of lead-rich residue.

TARNOWITZ PROCESS. This is the Carinthian method carried out in large furnaces. There is small loss by volatilization, low fuel consumption and moderate cost for labor. It yields a pure lead, but leaves a large residue in rich lead which must be treated in special furnaces.

FRENCH OR BRITTANY PROCESS. The work is done in large furnaces, large charges and a long roasting period, whereby considerable lead oxide and lead sulphate are formed, which necessitates the addition of fine coal during the reaction smelting.

This process has no advantages over those mentioned above, but on the contrary a high loss of lead is entailed and the life of the furnace is short.

HEARTH PROCESS. The ore, with fluxes and fuel (charcoal), floats upon the lead in a crucible-like hearth, three walls of which are made high. Through the back wall, air is blown into the charge above the molten lead.

The plant is simple and quickly constructed, but the

process entails a high loss of lead by volatilization, and the laborers are endangered by lead poisoning.

**Roast-reduction Process.** By this method the ore, after a long-continued roast, is melted with reducing agents to produce lead.

**Roasting.** Qualitatively, the reactions are the same as during the roasting period of the roast-reaction process. Quantitatively, however, the products are quite different, as it is not intended here to leave sulphur compounds in the roasted product for the purpose of reducing the oxides that are produced in the roast. The sulphides are, therefore, roasted down to the smallest amount necessary for forming matte (i.e., if the ores contain copper). In the absence of metals which are to be concentrated into matte, sulphates also are not desired in the roasted ore. The sulphates resulting from the roast are reduced during the reducing smelt. This re-formation of sulphides is at least superfluous; in the older processes the sulphate was decomposed by silica at the close of the roasting:

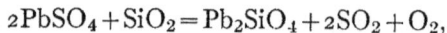

$$2PbSO_4 + SiO_2 = Pb_2SiO_4 + 2SO_2 + O_2,$$

and the temperature was raised correspondingly. Apparatus: reverberatory furnaces with hearths narrowing near the fire bridge.

The newer processes were suggested by the method of Huntington and Heberlein, in which a mixture of galena and lime is first roasted at about 700° C. and the roast product after being cooled to about 500° is blown to PbO in a converter.

Savelsberg dispensed with the preliminary roasting by moistening well the mixture of ore and lime and blowing it directly in a converter.

Finally Carmichael and Bradford treated a mixture of ore and gypsum directly in a converter.

In all three cases, the attempt is made to fulfil the conditions explained by Schenck and Rassbach for obtaining oxides, either directly or with the intermediate formation of lead sulphate, and to bring the ore to a proper condition for treatment in the subsequent reducing smelt. Plumbates form at the beginning as well as at the closing stages of these treatments.

The discovery of Doeltz that $CaSO_4$ does not react with PbS does not affect the Carmichael-Bradford process. In the plant operated in Australia the reacting substances are not $CaSO_4 + PbS$, but $CaSO_4 + SiO_2 + PbS$.

**Reduction Smelting.** In the roast-reaction process, the undecomposed PbS serves as a reducing agent for the oxides and sulphates formed in the roast. In this process, however, the reduction is effected by carbon and carbon monoxide:

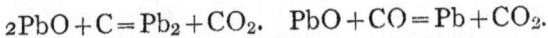

$$2PbO + C = Pb_2 + CO_2. \quad PbO + CO = Pb + CO_2.$$

The direct reduction of lead silicate in the original ore, or of that produced in roasting, cannot be effected by carbon; the lead oxide, therefore, in order to be acted upon by the reducing agents must be set free from the silicate by fluxes which also form an easily fusible slag:

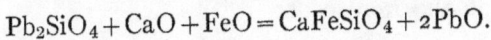

$$Pb_2SiO_4 + CaO + FeO = CaFeSiO_4 + 2PbO.$$

The roast reduction process is applicable to nearly all lead ores; if the silica is low, a silicious flux may be required.

In contrast to the roast-reaction process, in which only one apparatus is used for both stages of the process, the roast reduction process requires two forms of apparatus. Usually the

ROASTING is carried on in reverberatory furnaces or in funnel-shaped or bowl-shaped converters having tuyeres at the bottom.

THE REDUCING SMELT is always carried out in shaft furnaces.

**The Precipitation Process.** In this process the roasting of the sulphide ores as such is avoided, because the roasting is always difficult to complete on account of the fusibility of galena, lead oxide, and the other lead compounds that result from the roast. The sulphide ores, therefore, are fused directly with fluxes containing metallic iron:

$$PbS + Fe = Pb + FeS.$$

The iron sulphide resulting from this reaction exerts a strong solvent action on the other sulphides. If this fact were not taken into consideration, the first iron sulphide produced in the reaction would unite with the galena and thus prevent its decomposition by iron. Hence an excess of galena is used over the quantity required by the above equation:

$$xPbS + Fe = FeS(PbS)_{x-1} + Pb$$

This production of lead matte has several advantages:

THE MATTE may be roasted in low shaft furnaces, producing concentrated gases which may be utilized for making sulphuric acid. This advantage is not possessed by the roasting apparatus used in the roast-reaction and roast-reduction processes. With the roasted lead matte, a large part of the iron used in the original fusion is returned to the process, as this product consists chiefly of lead oxide and iron oxide:

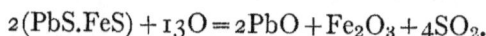

$$2(PbS.FeS) + 13O = 2PbO + Fe_2O_3 + 4SO_2.$$

The precipitation process, because of the above facts, may be used to advantage on ores which do not contain large quantities of other sulphides.

The constitution of lead matte has been cleared up by the investigations of H. Weidtmann, who found that no chemical compounds exist between PbS and FeS, but a eutectic exists having the approximate composition PbS + FeS and a melting point of 780° C. (1435° F.).

## Apparatus for the Roast-reduction and Precipitation Processes:

### Roasting Apparatus:

HEAP ROASTING is used as a preliminary operation for ores high in zinc and pyrite, the object being to sulphatize and leach out the zinc.

KILNS for roasting matte (see Copper).

REVERBERATORY FURNACES. In the early methods, the
long-hearth, hand reverberatory was used almost universally. The hearths were often up to 85 ft. in length and
from 8 to 16 ft. in width. If floor space was lacking, furnaces
were constructed having two hearths, one above the other,
each from 40 to 50 ft. long. At the sides of the hearths were
working doors, 5 to 6½ ft. apart. The reason for building such long hearths was because of the nature of the raw
material and of the roasted products. They are readily
fusible and require, therefore, a low temperature and a
correspondingly long time for the roasting. In spite of
the endeavor to pass the ore quickly through the furnace,
a considerable quantity remains continually on the hearth,
and is moved from time to time in small parcels toward
the firebox, where the final roasting takes place.

The construction of the hearth depends upon the special
object of the roast.

SLAG ROASTING. By this method the reaction between
$PbSO_4$ and $SiO_2$, mentioned above, is carried out as completely as possible. The portion of hearth lying nearest
the firebox is constructed in such a way and of such dimensions that the roasted product is fused here. The slagging
hearth, therefore, is made narrower and lower at this point
in order to concentrate the heat and collect the fused
material. At the side of the fusion hearth farthest from
the firebox, the hearth widens and usually becomes higher
in order that the temperature may be greatly lowered
for the roasting (Figs. 140–141).

SINTER-ROASTING. In this method a complete desul-
phurization is not sought; the chief aim is to heat the pulverulent roasted product only enough to sinter it. The
furnaces required for this work have a short sinter-hearth
near the fire-bridge, to which is joined the roasting-hearth
proper a step higher, making, as it were, a terrace. The
sinter-hearth does not need to be narrowed.

NON-SINTER ROASTING. In this work the product of roasting remains in a pulverulent state. The hearth of the furnace used for this method is a slightly inclined plane, sloping away from the flue-bridge toward the fire-bridge.

FOR REVERBERATORY FURNACES WITH STATIONARY HEARTHS many mechanical stirring devices have been used. These have been described under Copper, pp. 63 to 71.

REVERBERATORY FURNACES WITH MOVABLE HEARTHS are used in the Huntington-Heberlein process, espe-

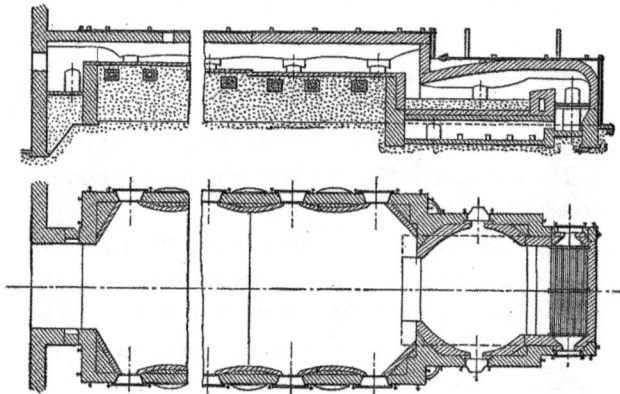

FIGS. 140-141.—Reverberatory Furnace with Slagging Hearth.

cially those with revolving circular hearths, having stationary rabbles, similar to the heating furnaces used in the manufacture of coal briquettes.

CONVERTERS are used in the following processes:

HUNTINGTON-HEBERLEIN: conical, cast-iron vessels 5-8 feet wide at the top and 5-6½ ft. deep, fitted at the bottom with grates to support the charge and to distribute the blast, which is also admitted at the bottom. Capacity 5-8 tons.

CARMICHAEL-BRADFORD: conical, cast iron vessels; upper diameter 6 ft., lower diameter 4 ft., depth 5 ft.

Above the bottom, which is arched downward, is placed
a perforated shell arched upward which acts as a grate.
The blast is admitted into the intermediate space thus
formed.  Capacity, 4 tons (Fig. 142).

SAVELSBERG.  Nearly hemispherical, cast-iron pots about
6½ feet in diameter.  Air admitted from below.  Capacity
8 tons.  Blast required, 250 cu.ft. per min.  Blast pressure
at start, 4 to 8 in. of water, later 20 to 24 in.  Time
of blowing, 18 hours (Fig. 143).

**Apparatus for the Reduction and Smelting:**

SHAFT FURNACES.  Of the older constructions the low

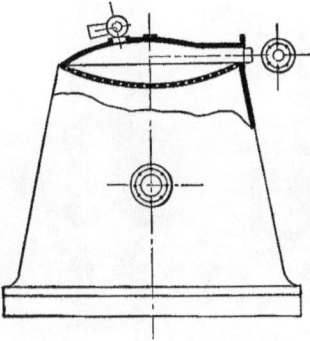

FIG. 142.—Carmichael-Bradford        FIG. 143.—Savelsberg Converter.
Converter.  Scale, 1 : 50.                 Scale, 1 : 50.

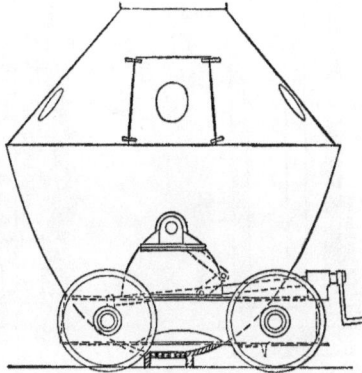

furnaces (Krummöfen) have almost wholly disappeared,
while the medium furnaces (Halbhochöfen) are found only
in a few plants.  The high furnaces (Hochöfen) are used
almost exclusively in modern plants.  They are, of course,
much smaller than the blast furnaces used in the iron
industry.

The "Krummöfen" are low shaft furnaces not more
than 6½ ft. in height, with square, rectangular or trapezoidal
cross-section.  They usually have only one tuyere which is
placed in the back wall.  Various forms of these furnaces

are described in the oldest metallurgical literature (*"Agricola de rebus metallicis,"* 1657, pp. 313–319; Schlüter, *"Hüttenwerke,"* 1738, tables XX–XXX).

"HALBHOCHÖFEN." These are low shaft furnaces, 7–14 ft. in height from the tuyere level to the throat. They are usually of trapezoidal cross-section with from 2 to 8 tuyeres on one side of the trapezoid. Of these furnaces the Stolberg furnace has been most widely adopted and was

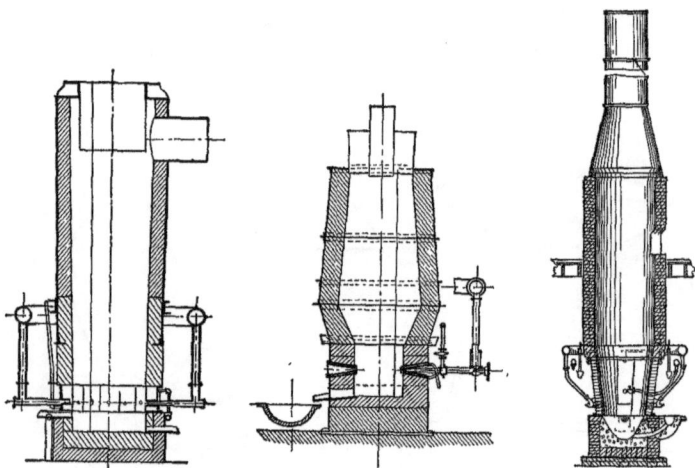

FIG. 144.—Pilz-Freiberg.     FIG. 145.—Upper Harz.     FIG. 146.—Allis-Chalmers.

also used until recently in Freiburg. They are still in operation at the lead plants in Stolberg and Mechernich.

"HOCHÖFEN." Although the name is usually applied to blast furnaces having at least 13 ft. between the tuyere level and throat, shorter furnaces of this type are found in some lead plants.

Modern blast furnaces are built with circular and rectangular horizontal cross-sections (Figs. 144–147). The shaft is usually made of brick and the bosh and smelting zone are water-jacketed. The throat is open, except during blowing in and blowing out, when it is covered with a

hood.   Downcomers start from the centre of the throat
or from the side under the feed floor.   They are usually

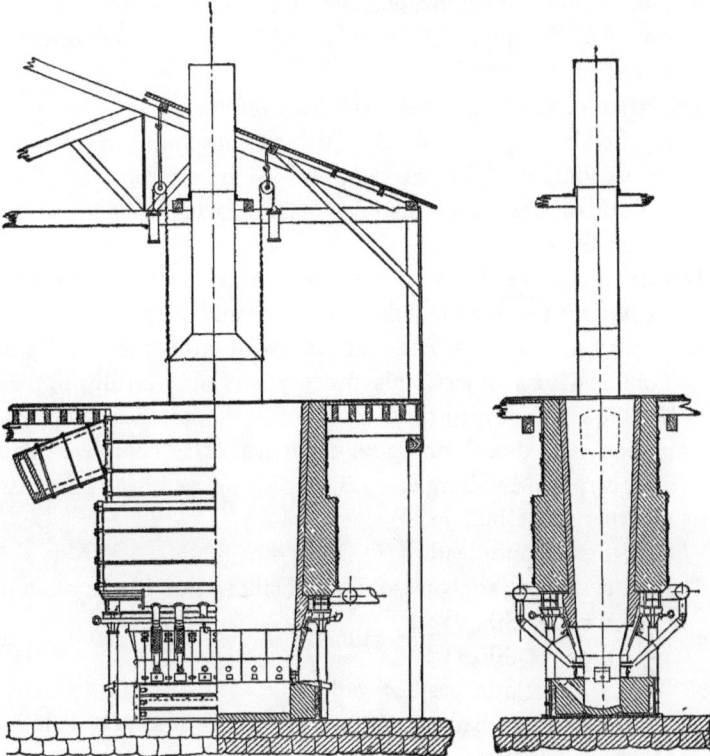

FIG. 147.—North American Water-jacket Furnace.   Scale, 1 : 150.

made of encased sheet-iron pipes.   Fore-hearth: crucible
furnace or Arent's siphon tap.

The direct electric smelting of lead ores has not yet been
successful.

## (C)  Lead Refining

The crude lead obtained in the preceding processes may contain various quantities of all the metals found in the ores, also
metallic compounds such as sulphides, arsenides, antimonides,
etc.   Among the impurities are:  Au, Ag, Cu, Sn, Sb, As, Bi, Co,

Ni, Fe, Zn, and S. They may be separated from the lead by the following operations:

**Liquation and Crystallization Processes** (see Silver). In this way Au, Ag, and Cu are removed by being concentrated in a part of the lead.

**Oxidation of the Impurities** (true lead refining).

(*a*) Oxidation of S, As, Sb, Bi by atmospheric oxygen.

(*b*) Oxidation of Zn, Fe, Co, and Ni by steam.

' Both of the above methods have already been described under Silver.

**Oxidation of the Lead,** the so-called *cupellation process* for recovering precious metals (see Silver and Bismuth).

**Electrolysis.** After several unsuccessful attempts by Keith, Tommasi and others, this process was successfully applied by Betts, in 1907, to the refining of antimonial lead and base bullion. A description and sketches of a complete plant are given in Metallurgie, 1908, **5,** 68.

ANODES: base bullion.

CATHODES: pure lead.

ELECTROLYTE: aqueous solution of silicon fluorides containing

22.7 oz. $PbSiF_6$ ⎫
9.4 oz. $H_2SiF_6$ ⎬ per gallon.

6.7 oz. gelatin per 100 gallons.

TEMPERATURE: between 17° and 57° C. (63 and 135° F.)

CURRENT DENSITY: 10 to 12 amp. per square foot of cathode.

POTENTIAL: 0.15–0.36 volt.

## Properties of Lead:

SPECIFIC GRAVITY: 11.4.

COLOR: bluish gray, shiny.

MECHANICAL PROPERTIES: low tensile strength, high ductility.

STRUCTURE. Ingot fracture (Fig. 148). Crystallized shell from which the molten metal was removed before freezing was completed (Fig. 149). Granular structure obtained by etching the bottom surface of a lead ingot (Fig. 150). '

MELTING POINT: 327° C. (621° F.)

VAPORIZATION takes place at red heat, though the boiling
point is between 1200 and 1300° C. (2200 and 2375° F.)

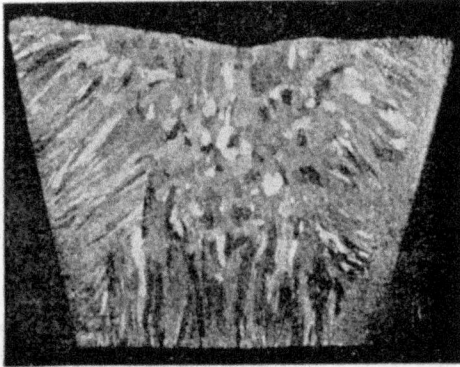

FIG. 148.

FIG. 149

THERMAL AND ELECTRICAL CONDUCTIVITY: low, the latter
0.0756 that of Ag.

ALLOYS with most metals. Its most important metallurgical
alloys are those with the precious metals, for which it is

utilized as a solvent: with zinc (limited solubility), with antimony (hard lead) and with tin (solder).

CHEMICAL BEHAVIOR: readily oxidized on the surface by O, $H_2O$, and $CO_2$ of the atmosphere. The coating of oxide and basic carbonate thus formed prevents further corrosion.

In the same way the action of HCl and $H_2SO_4$ is restricted even in the presence of oxygen. Without the presence

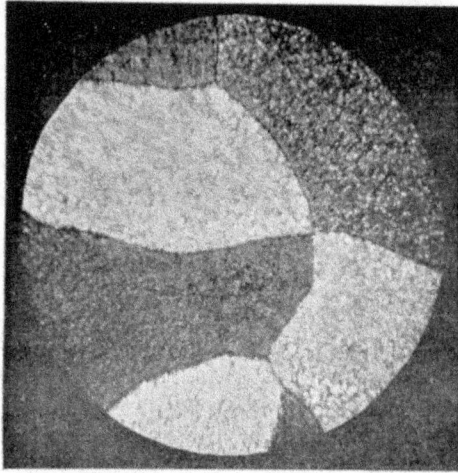

FIG. 150 ($\times$33).

of oxygen or oxidizing agents in the acid, it is hard to dissolve lead in acids, because the electrolytic potential is only +.151 against H.

There are two common oxides, PbO and $PbO_2$, the former basic, the latter acid. These unite to form a plumbate, $Pb_2PbO_4$ (which, together with free PbO, constitutes red lead).

Of free sulphides, only PbS is known; it dissolves readily in other sulphides (matte), but forms few chemical compounds.

# TIN

## Sources

**Natural Sources:**

OXIDIZED ORES.

Tin-stone, or cassiterite, $SnO_2$. In primary deposits, lode tin; in secondary deposits, stream tin. Gangue: acid eruptive rocks, particularly granite, also quartz, $CaF_2$, $FeWO_4$, $CaWO_4$, sulphides such as $Cu_2S$, $PbS$, $Bi_2S_3$, $MoS_3$, etc., and arsenides such as $FeAsS$.

SULPHIDE ORES.

Tin pyrites, $FeCu_2.SnS_4$, which is much rarer than tin-stone.

**Other Sources:**

METALLIC INTERMEDIATE PRODUCTS AND BY-PRODUCTS:

Hard Head and Liquation Dross (Fe-Sn alloy).

Tin plate clippings (Fe covered with a coating of Sn).

White Metal in the form of impure turnings: Sn, Pb, Cu, Zn, Sb.

OXIDIZED BY-PRODUCTS:

Slags: Silicates and Stannates.

Ashes: Oxides mixed with metallic grains.

## (A) Concentration

**Mechanical Concentration.** On account of the high specific gravity of tin-stone, viz., 6.8 to 7, the mechanical concentration of tin ore is carried out quite commonly, and sometimes by the simplest means.

**Chemical Concentration.** Among the minerals accompanying tin-stone are found, as mentioned above, sulphides, arsenides and tungstates. The first two are almost always present, but of

133

the last mentioned only a few ores contain enough to cause trouble.

**The removal of S and As** is effected by an oxidizing roast in a simple roasting furnace, usually fed by hand. For ores with much sulphide of Fe and Cu, it has been proposed to carry out the roasting in such a way that the sulphates of these metals are formed, which can then be leached out, but such a process does not seem to have met with much practical application.

**The removal of tungstates** is necessary with ores containing much tungsten, because tungstates favor the passage of tin into the slag. Even a small percentage of tungstate suffices not only to carry over considerable amounts of tin into the slag on smelting the ore, but also hinders the reduction of tin in the subsequent smelting of the slag. Hence, even in smelters that work without previous concentration, with ores containing small amounts of tungsten, it is worth while to take the slags in which considerable tin has accumulated, and remove the tungsten in order to recover the tin by another smelting. The removal of tungsten is effected by an oxidizing roast of the pulverized ore or slag, with sufficient $Na_2CO_3$ to convert the W into $Na_2WO_4$. The latter is removed by lixiviation, whereupon the tin can be obtained without difficulty. Fuller details concerning such work will be given under Tungsten.

# (B)  Extraction

Inasmuch as tin-stone is the principal raw material, which, when necessary, is concentrated or purified by the above-mentioned work, the chief operation of the tin smelter is of a comparatively simple nature. It consists of a reducing fusion of the ore, and other oxidized waste products, and the subsequent working over of the slag. During the last two decades it has also been found possible by an electrolytic process to recover tin from the scrap obtained in the preparation of tin plate.

**The Smelting Operations for Oxidized Ores and Waste Products** are the following.

1. **Reducing Fusion.** Concerning the reactions that take place during this process, singular views are to be found in metallurgical literature. The ready reducibility of $SnO_2$ by KCN, which is preferred as a flux and reducing agent in making the blow-pipe test for tin, has led to the assumption that during smelting the carbon of the coal used for effecting the reduction is partly converted into cyanide, which then acts as in the blow-pipe test. There has never been any proof of this and, moreover, the assumption is very improbable, because, in most cases, the slag is not kept basic enough nor hot enough for the formation of cyanide. An experimental investigation by Mattonet, in the laboratory of the author, has shown that the formation of a singulo-silicate slag is most favorable for the extraction of the tin and for the slagging off of the other metals.

According to the nature of the ore, the fuel, and the other working conditions, it has been customary to make use of either a blast furnace or a reverberatory furnace for smelting. The above-mentioned researches of Mattonet, however, have established the fact that an electric furnace can be of service for producing slags low in tin. In both blast-furnace and reverberatory smelting, the extraction of the tin is made difficult by the unfavorable position of the reduction temperature (1000° to 1100° C.) in respect to the melting point (232°) and the boiling-point of the metal. The latter, to be sure, is said to be above 2100°, but volatilization of the metal takes place at temperatures far below the boiling point. The furnace charge must, therefore, be heated to 1100°, and the combustion gases, which furnish the source of the heat and its transference, must be kept at least 200° hotter, i.e., at about 1300°, so that the Sn is heated to about 1000° above its melting-point, at which temperature it shows a tendency to volatilize with the reaction-gases and also possesses a high solution tension which favors its passing into the slag. In

the electric process, particularly since the charge itself can
take part in the heat transference, the surroundings are
kept colder than the charge.

THE BLAST-FURNACE PROCESS, which is the oldest method
of working, is applicable if a very pure, lumpy ore,
and a pure fuel, e.g., wood charcoal, are available. The
smelting takes place sometimes without any additions
of flux, and sometimes with slags obtained from tin ores,
used in amounts varying from 25–50% of the weight of ore.
In the Chinese and Malay tin districts, the work, in some
cases, is still carried out with very primitive apparatus.
Whereas the original blast-furnaces of the Sunda Islands
consisted of pits, about 20 in. deep and 12 to 16 in.
wide, dug in the earth and provided with hand-driven
blast which passed through hollow tree trunks, the Chinese
draft and blast furnaces are prepared by tamping
clay within a barrel-shaped form made of bamboo
rods and then carving out the shaft and the holes for
blowing and tapping. These furnaces are built up to
6 ft. in height, with a shaft up to 16 in. wide and about
5 ft. 3 in. high. Such furnaces, when used for smelting
the ore, will work with natural draft, provided the material is sufficiently coarse; and those used for smelting the
slag work with a blast, which, as in the above instance, is
produced by hand, using hollow tree trunks that are provided with wooden pistons. (Fig. 151.) Even the larger
blast furnaces, such as are used in India at works conducted by Europeans, as well as in Bohemia and in Alten-
berg, are built so short and narrow that they can almost
be classed as "dwarf furnaces." Concerning the construction of such furnaces, enough was given under
Copper and Lead, so that it is unnecessary to go into
further details here.

The method of smelting tin ores which is in common use
to-day is by means of the REVERBERATORY FURNACE,
because the ore is usually fine grained in nature and of

low grade, so that the use of the blast furnace is out of the question.   Although formerly considerable importance was attached to the nature of the fuel, inasmuch as only those varieties of coal are suitable which produce long flames, to-day this difficulty can be overcome by the use of gas firing.   Even in this case the furnace is almost always charged merely with ore and coal; when the gangue is too acid it is sometimes necessary to add limestone and small

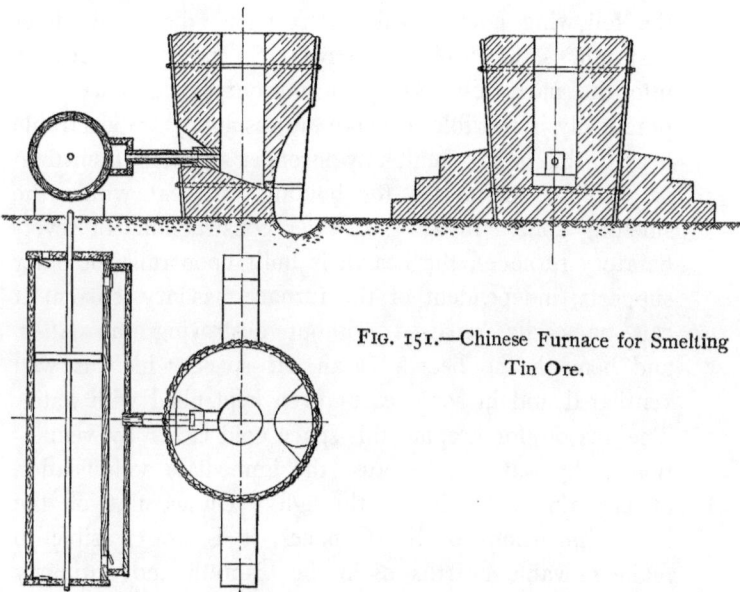

FIG. 151.—Chinese Furnace for Smelting Tin Ore.

amounts of fluorspar.   In order to prevent mechanical losses of the powder, the mixture is usually moistened with water.   After being introduced into the hot furnace, the charge is levelled off and heated strongly for from 5 to 8 hours with closed doors and nearly checked draft; after the mass has softened, it is strongly heated for from 45 minutes to an hour, the metal is tapped and the slag, which solidifies readily on cooling, drawn off.   In other works, the slag is thickened by working in some coal so that it can be removed through the slag door before tapping the metal, which is then covered with only little slag. According to the size of the reverberatory furnace, the amount of ore used in a single charge varies from 1600 lbs. to 4 tons, the time of heating from 6 to 12 hours, the consumption of fuel from 60 to 120% of the ore, the labor employed from 1 to 3 men per furnace, and the smelting losses up to 7% Sn.

As regards the construction of the reverberatory furnaces, the following points, which were partly discussed under Bismuth, should be considered: Sn penetrates readily into the narrowest fissures of masonry, and since it is practically impossible to make an absolutely impenetrable hearth, this is not built any stronger at the bottom than is absolutely necessary for holding the heat within the smelting zone. In all the newer constructions of reverberatory furnaces, the hearth is built upon rails, or other supports, independent of the furnace-masonry, so that it can be easily repaired without destroying the latter, and beneath the hearth is an air space which is well ventilated and in some cases even kept filled with water. The reason for keeping this space cold either by ventilation or by water, is to cause the immediate solidification of any tin that trickles through. On account of the large dimensions of the furnaces, it is not feasible to make movable hearths, as in the bismuth and antimony industries (cf. p. 114, Figs. 134–137). It is necessary, however, to have the fireplace and flue built in such a way that they are separated by air spaces, or some insulating material, from the parts of the furnace containing the hearth. Otherwise the molten tin would tend to run toward the fire bridge or flue bridge and find a particularly favorable opportunity to penetrate through them.

The smelting of tin in an electric furnace has been carefully studied by Mattonet, in the author's laboratory. Experiments performed in the attempt to smelt impure tin ores, without a preliminary roasting, in such a way that the Cu

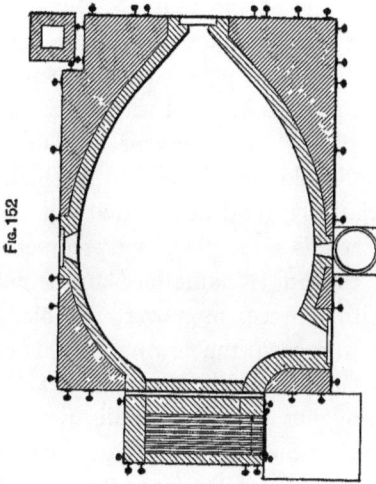

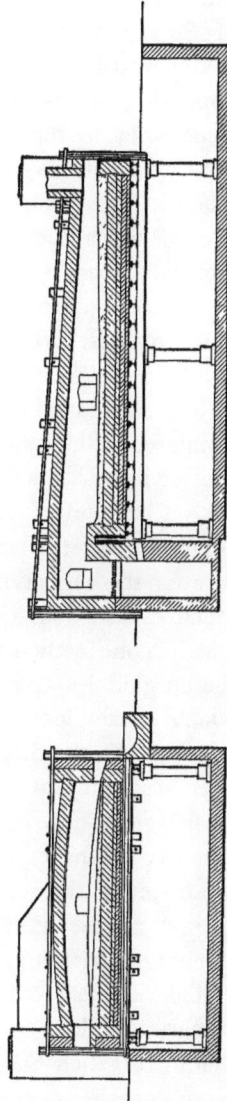

Fig. 152

Reverberatory Furnace for Tin.

Fig. 154.

Fig. 153.

and a part of the Fe would be converted into a matte and the Pb into an acid slag, proved unsuccessful. Even the sulphur in the ore passed to a considerable extent into the crude tin, which contained up to 2.7% S. The yield of Sn under these conditions was also unsatisfactory. Better results were obtained with roasted ores and an approximately neutral slag, in which, however, by adding some soda to the charge the melting-point was lowered somewhat and $Na_2O$ made a part of the basic constituents. Since at most only 450 lbs. of $Na_2CO_3$ (cost \$2.50) are needed for a ton of Sn, the cost of working is not increased, because more than enough Sn is prevented from going into the slag to compensate. The slag obtained was perfectly free from Sn globules (in the case of other methods of smelting it is permeated by pellets of Sn), and contains only from 1 to 1.5% of slagged Sn, as compared with from 2 to 5% by other processes. Direct heating by resistance is applicable for this method of smelting on account of the relatively good conductivity of the charge and of the slag. In this way, the proper temperature for the reduction of the Sn is maintained and is not much exceeded in the bath. Since, moreover, the blast, which in one method tends to prevent the flowing together of the Sn globules, is absent and an acid lining of the hearth, which causes losses by slagging and by volatilization, is unnecessary, the direct yield of Sn is so good that, whereas it is absolutely necessary to recover Sn from the slag in smelting by the blast furnace or by the reverberatory furnace, in the case of electric smelting this subsequent treatment of the slag can be dispensed with altogether or limited at the most to one smelting. Consequently the smelting of rich and poor slags, as described below under 2 and 3, refers only to blast-furnace and reverberatory furnace practice.

2. **Smelting of Rich Slags.** The slags obtained by smelting the ore contain, as mentioned above, tin which is mechanically enclosed, as well as that which is slagged. As soon as a sufficient amount has accumulated, it is smelted,

usually in a reverberatory furnace, together with other metallic waste (dross from the refining, and the "hardhead" or residue which remains on the hearth), and, in case not enough iron is present to cause the separation of the slagged tin, some scrap iron is added also. Obviously it is necessary to conduct the work with a reducing flame and with some fuel added to the charge.

Since the smelting of rich slags requires a higher temperature than that of ores, this process is often carried out for the purpose of seasoning new hearths which are to be used for smelting ore. A considerable part of the slag adheres to the masonry of the hearth and either remains solid during subsequent smelting of ores, or else is so viscous that metal does not flow through it.

3. **Smelting of Poor Slags.** This is carried out in many works in blast furnaces, usually in small water-jacketed furnaces having a circular cross-section; large pieces of coal or coke are mixed with the charge.

Although a fairly pure tin is obtained directly by the two former operations (the crude metal from rich slags even contains 95% Sn), in this case the crude metal is a fairly rich iron alloy (80% of Sn, 20% of Fe).

**Electrolytic Solution and Deposition.** For twenty years this process has been carried out in working up tin-plate waste. The object, in most cases, is to recover the tin from old scrap made in the manufacture of tin-plate. According to the thickness of the sheet iron which was given a coating of Sn, the material contains from 3 to 5% of Sn. As a rule, however, it is not safe to reckon on more than 2 or 3%. The recovery of the tin has been carried out most successfully in the factory of T. Goldschmidt at Essen. Most other works have made use of the same method and profited by the experience gained there. The apparatus and conditions are as follows:

ELECTROLYZING VESSELS: iron tanks. These are connected with the circuit and serve as cathodes.

ANODES: the tin-plate clippings packed not too firmly in a basket made of coarse iron wire (about 100 lbs. of clippings are placed in a basket).

CATHODES: sheet iron and the walls of the tank.

ELECTROLYTE: this consists at first of a caustic soda solution containing at the most 9% of NaOH, corresponding to 7% $Na_2O$. During the electrolysis the caustic soda is partly converted into carbonate and partly into stannate. Of the original 7% of $Na_2O$, 3 to 3.5% remains as the free hydroxide while 1 to 1.5% of $Na_2O$ is converted into stannate and 1.7 to 2.8% of $Na_2O$ unites with $CO_2$. If this amount of carbonate is exceeded, the electrolyte

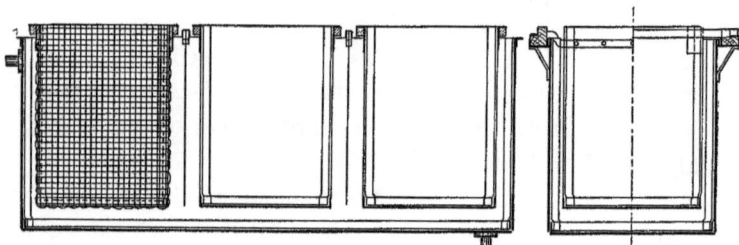

FIG. 155.                    FIG. 156.

must be worked over, by treating with $Ca(OH)_2$, whereby $CaCO_3$ is precipitated and NaOH formed.

TEMPERATURE: 70° C. (158° F.).

POTENTIAL: 1.5 volts.

Concerning the reactions that take place during the electrolysis there is but little reliable information. It is known that the Sn goes into solution as stannate, and therefore forms a complex anion, and that it is consequently indirectly precipitated (by Na?).

PRODUCTS:

ELECTROLYTIC TIN. Inasmuch as old tin-ware that is worked up in this way contains more or less solder, the Sn obtained by electrolysis usually contains some Pb. For this reason, the Sn is deposited in the form of crystalline grains and does not form coherent plates. On being dried it becomes covered with a layer of oxide, so that it must be subjected to a reducing smelt.

LEAD CARBONATE: the first precipitation product in working up impure electrolytes.

STANNIC ACID, or $SnO_2$: the second precipitation product from impure electrolytes.

## (C) Tin Refining

It is clear, from what has been said concerning the extraction of tin, that the principal impurity of the crude metal is iron, and that to a less extent metals such as lead, copper, tungsten, etc., may be present.

1. **Liquation.** By this process it is chiefly the Fe which is removed either in the free state, or in the form of a rich Fe alloy. The impure tin is carefully heated so that the melting-point of tin is exceeded as little as possible, or, in other words, so that the tin alone is liquefied, and the above-mentioned Sn-Fe alloy remains behind as a solid. For carrying out the process it is customary to make use of small, simple reverberatory furnaces, the sloping hearths of which are divided by steps into two or three sections. When there are three such sections the crude tin is placed in the middle one; the pure tin that liquates out is allowed to solidify and is then transferred into the third section, which is the farthest away from the fire-bridge. The tin that flows away from here is purified further if necessary. The residue remaining in the third compartment, the liquation dross or "hard-head," goes back into the second section, and all the residues from this in turn, including that from the crude tin, are placed in the first section, which lies nearest to the fire. The tin that flows away from the first compartment, after solidifying, is placed in the second, just as that from the second goes to the third. The dross from the first compartment is used partly as an iron-bearing flux in smelting

tin slags, and partly for preparing Pb-Sn alloys, by simply melting it up with Pb. In the smelter certain standard alloys are prepared and sold as such, e.g., that with 50% Sn and 50% Pb) (common solder).

2. **Oxidizing Fusion.** Although the bulk of the Fe, W and a part of the Cu is removed by the liquation process, there still remain small amounts of these impurities in the tin. They may be removed with the aid of atmospheric oxygen. To accelerate the oxidation process, there is placed in the iron kettle which contains the molten tin, a pole of green wood from which vapors of water and other decomposition products of the wood are at once evolved. The melted metal is stirred up by the escape of these gases, so that new portions of it are constantly brought to the surface, where they come in contact with the oxygen of the air. The water vapor given off by the pole also tends to assist in the oxidation of the impurities.

3. **Crystallization of Tin.** When Sn is strongly contaminated with Pb it is almost impossible to effect a sharp and complete separation of the two metals. To be sure there has been no want of attempts in this direction. Since, however, considerable quantities of Pb-Sn alloys are used technically, it is not at all necessary in most cases to attempt such a separation—it suffices to enrich the metal. This, as the experiments conducted by the author with the aid of Mattonet have shown, can be carried out in much the same way as in the separation of lead and silver, by the crystallizing out of the lead. From the freezing-point curve (Fig. 157) it is evident that the two metals are soluble in one another in all proportions and that there is a eutectic formed with 37% Pb and 63% Sn which melts at about 180° C. If, therefore, crude tin containing a small percentage of Pb is allowed to cool slowly, it is obvious that pure Sn will crystallize out, leaving behind a mother metal which constantly increases in Pb content, until finally the composition of the latter is that of the eutectic alloy. In

practical work, the cooling will not be carried out as far
as this, but will be stopped while the alloy is still liquid
and contains somewhat less Pb than the eutectic does.
This is drawn off in the same way as in the desilverization
of crude lead, leaving behind an alloy containing a little Pb.
Naturally, the process does not yield perfectly pure tin
the first time, simply because the Sn crystals are more or
less contaminated with the mother metal, which contains Pb
and adheres to some extent to the crystals.  Thus, to pre-

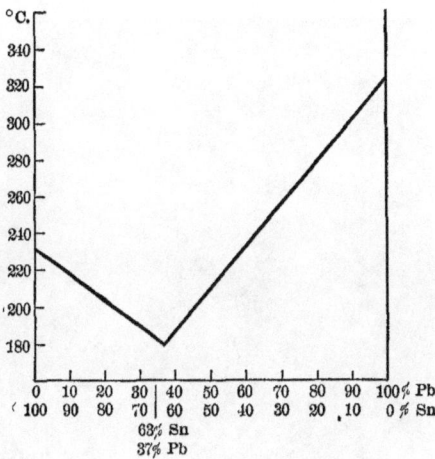

Fig. 157.

pare perfectly pure Sn it is necessary to repeat the process a number of times, and, as in the desilverization of
lead bullion, there is no difficulty in the work. The Sn
alloy obtained by this process of separation need only be
brought to a content of 10% Pb, which is used very extensively
as such for making terne-plate.

**Properties of Tin:**

SPECIFIC GRAVITY: 7.3.

COLOR: white with a pale bluish tinge; yellow when hot.

MECHANICAL PROPERTIES: low tenacity and hardness,
    high ductility; most ductile at about 100°; at 200°

so brittle that it can be pulverized. At low temperatures
it becomes changed into gray tin powder.

STRUCTURE: see Figs. 158, 159, 160.

MELTING-POINT: 232° C. (450° F.).

BOILING-POINT: 2100–2200° C. (3800–4000° F.); volatilizes
perceptibly at 1200° C. (2200° F.).

THERMAL AND ELECTRICAL CONDUCTIVITY: good, the
latter about 0.13 that of Ag.

ALLOYS with almost all metals. Many Sn alloys are valued

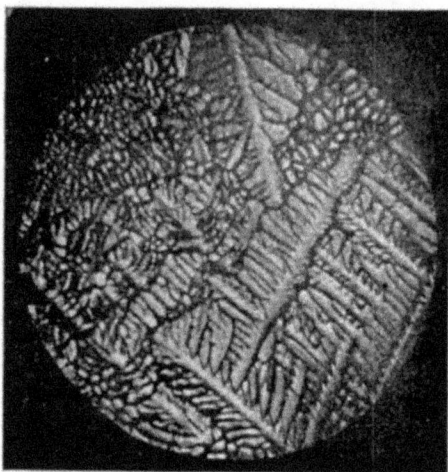

FIG. 158.—Surface of Impure Tin (×33).

highly (with Cu = bronzes; with Cu, Pb, Sb = bearing
metal, and solder; with Sb, Ni, W, etc., metal for fine-
art castings). In experiments for obtaining alkali and
alkaline-earth metals by the electrolysis of their fused
salts, Sn has been found to be a particularly good solvent
for these metals.

CHEMICAL BEHAVIOR: very stable toward O and $H_2O$
at temperatures below its melting-point; at higher
temperatures it oxidizes readily and also combines
readily with S, P, As, and Sb. Cl and other halogens

act upon it even at ordinary temperatures, forming $SnCl_4$, etc. Of the mineral acids HCl dissolves the metal best, $H_2SO_4$ only slowly, and $HNO_3$ converts

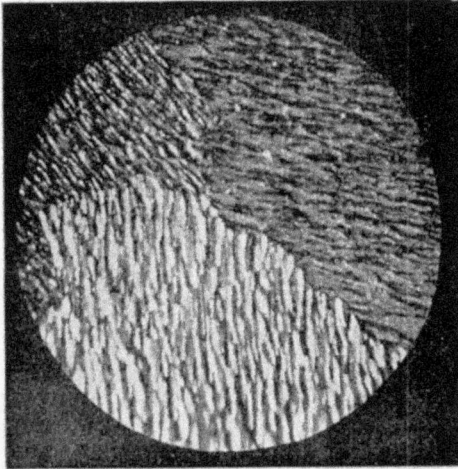

FIG. 159.—Deeply Etched Under-surface of a Cast Tin Ingot.

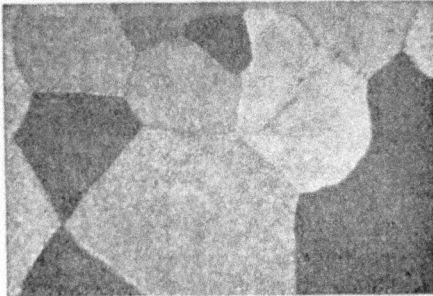

FIG. 160.—Large Grains caused by Tempering Tin.

it into insoluble metastannic acid. which is probably a polymer of $H_2SnO_3$. This dissolves in alkali to form salts soluble in water, called stannates. With O, S and on dissolving in acid, two series of compounds are formed, those of the stannous oxide type (SnO) called the stannous compounds (Sn + +) and those of the stannic oxide type (SnO$_2$) called stannic compounds (Sn + + + +). In the presence of strong bases Sn loses its basic properties and unites with O to form an anion (stannites and stannates). In acid aqueous solutions, as well as in fusions, Sn is precipitated by Pb whereas in basic solutions it precipitates Pb.

# ANTIMONY

## Sources

**Natural Sources:**

NATIVE: occurs as such rarely.

As SULPHIDE: $Sb_2S_3$, which is known as stibnite; this is the most important ore of antimony.

As OXIDE: $Sb_2O_3$, called senarmontite or valentinite.

**Other Sources:**

Waste products from the working up for other metals of ores containing antimony, particularly in the smelting of lead (dross).

## (A) Concentration

**Liquation.** The ready fusibility of $Sb_2S_3$ makes it possible to melt out this compound from its ores, leaving behind, to be sure, a residue rich in Sb, since a considerable amount of $Sb_2S_3$ (up to 20%) entangled in the gangue remains behind in the liquating apparatus. In spite of this difficulty, such work is still carried out to a considerable extent in Hungarian, Japanese, and Chinese antimony districts, because the concentration product, the so-called *antimonium crudum*, is still much used in English antimony works for the preparation of pure antimony, and in German chemical factories for the manufacture of antimony preparations. If it were desired to-day to erect antimony works in the vicinity of the mines, naturally the liquation process would not be used at all, because the residues would have to be worked over considerably on account of the large amount of antimony they contain, and because there is not a large amount of gangue to slag off when the ores are treated directly.

149

APPARATUS: Crucibles or Tubes placed either in air furnaces of the simplest type, or upon the hearth, or in a special heating chamber of a reverberatory furnace. The crucibles are provided with holes in the bottom and the tubes are open at both ends so that the melted $Sb_2S_3$ flows out into collecting vessels. Crucibles heated in a reverberatory furnace are placed over a tapping hole which leads from the bottom of the hearth to the outside. Reverberatory Furnaces with deep hearths (see Fig. 47, p. 29).

**Oxidizing Roast with Sublimation.** The ready volatility of $Sb_2O_3$, and the demand on the part of chemical industries for this oxide, has led to an extensive practice of subjecting antimony ores to an oxidizing roast, in spite of the fact that sulphide ores can be worked into metal without any preliminary roasting. Then again, ores can be worked with in this manner from which no antimony sulphide could be obtained by liquation and with which the expense of slagging off all the gangue would be too great on account of the large amount they contain. Now antimony on being roasted can be converted into $Sb_2O_3$, which has a basic character, or into $Sb_2O_5$, which is acidic in nature. The latter is formed more readily in the presence of basic substances, but as $Sb_2O_3$ is itself basic, considerable quantities of the higher oxide are likely to result unless special measures are taken to prevent its formation or to cause its reduction after it has once been formed. In all cases, the aim is to make $Sb_2O_3$. Chemical industries demand this oxide in a condition as free as possible from $Sb_2O_5$, because the former is readily soluble in acid and the latter is not. But in extracting the antimony from the ores by the oxidizing roast, it is necessary to maintain the conditions for the formation of the lower oxide because this is readily volatile, whereas $Sb_2O_5$ is not, and cannot, therefore, be removed as readily from the gangue by volatilization. If, on the other hand, the sublimation product is to be used for the manufacture of pure antimony, it is not necessary that it should be free from $Sb_2O_5$.

APPARATUS: Although, in metallurgical literature, a number of forms of apparatus and processes are described which are very simple, yet in practice only low blast-furnaces or reverberatory furnaces are employed. In both cases, it is, of course, necessary to provide arrangements for properly catching and condensing the volatilized oxide.

**Lixiviation.** It is well known that $Sb_2S_3$ is soluble in aqueous solutions of alkali sulphides forming sulphantimonites, or sulphantimonates upon the addition of free sulphur.

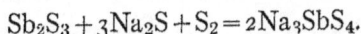

$$Sb_2S_3 + 3Na_2S = 2Na_3SbS_3,$$

$$Sb_2S_3 + 3Na_2S + S_2 = 2Na_3SbS_4.$$

Chemical manufacturers for a long time have made use of these reactions in preparing artificial, precipitated antimony sulphide. Usually solutions of $Na_3SbS_4$ are first prepared and an intimate mixture of $Sb_2S_3$ and $S_2$, which is regarded as $Sb_2S_5$, is obtained on acidifying. If, however, the solution is to be used for the electrolytic deposition of antimony, the addition of S during the lixiviation is not only unnecessary, but undesirable. A solution of $Na_3SbS_3$ can be electrolyzed directly, whereas one of $Na_3SbS_4$ must first be treated with an excess of $Na_2S$ or of $NaOH$.

APPARATUS: iron tanks, conical at the bottom with converging sides, and provided with pipes for introducing steam, which serves to heat the solution and to stir up the ore in it.

# (B) Extraction

**Reduction Methods.** For oxidized ores or metallurigcal products, the same methods are employed as in the case of the extraction of Bi (pp. 111 to 115) although in this case it is necessary to use cheaper fluxes because the price of Sb is to that of Bi as 1:20.

**Precipitation Methods.** This depends, as in the like-named processes for obtaining Pb and Bi, on causing the reaction

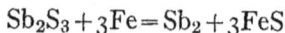

$$Sb_2S_3 + 3Fe = Sb_2 + 3FeS$$

to take place in the melted mass.

As regards the apparatus and method of conducting the operation, what was said under Bi holds in the main here also. Crucibles and reverberatory furnaces are used. In reports concerning precipitation in reverberatory furnaces a process is frequently described in metallurgical literature as the *English Process*, but which appears to be identical with the practice that has been followed everywhere that $Sb_2S_3$ has been melted with Fe. Under Bismuth it was pointed out that it is always necessary in melting charges containing easily volatile material, to provide a sufficient cover of fused material so that when a fresh charge is added it will at once sink below the surface. In smelting rich antimony ores, or the concentration product called *antimonium crudum*, it is clear that FeS will form the principal constituent of the slag. FeS, however, is known to be an excellent solvent for other sulphides and for metals, particularly Fe itself. Fe, moreover, is the agent that precipitates Sb from $Sb_2S_3$ in this process. Just as in an ordinary reducing fusion it is necessary to provide a reducing atmosphere, and solid reducing agent in the furnace and charge, so it is clear that care must be taken in the precipitation process to have enough Fe present in the furnace before the antimony ore is introduced. Alternately charging with Fe and $Sb_2S_3$ is prescribed, therefore, in order that the Fe may be given time to dissolve in the molten FeS, and thus attain a sufficiently fine state of division to act vigorously enough upon the $Sb_2S_3$ with which it subsequently comes in contact.

**Electrolysis.** The solution of $Na_3SbS_3$ obtained by extraction with alkali sulphide (cf. Concentration Work, above) can be electrolyzed directly, as has been shown elsewhere, without any additions whatever to the bath, such as are recommended in the analytical estimation of Sb by electrolysis. Even although the processes that take place have not been wholly explained, it has been shown that Sb is present in the anion of the above salt as well as in $Na_3SbS_4$. The separation of the Sb on the cathode, therefore, must be the result of a secondary reaction; first the Na ions are discharged and then the Sb is precipitated from the sulphide. This partly accounts for the fact that it is difficult to deposit the Sb upon the cathode in a dense condition. The total result of the electrolysis may be expressed by the equation,

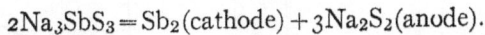

$$2Na_3SbS_3 = Sb_2(\text{cathode}) + 3Na_2S_2(\text{anode}).$$

In 1887 Borchers established the fact that large quantities of $Na_2S_2$ are formed at the anode, as well as some NaHS and $Na_3S_2O_3$. The last-mentioned compound is formed readily by the oxidation of $Na_2S$ or of NaHS, and this explains why $Na_2S_2O_3$ can be obtained by blowing air into solutions from which antimony has been deposited electrolytically. The crystallized salt is a common commercial article.

Special conditions for the electrolysis are:

ELECTROLYZING VESSELS: iron tanks with rectangular cross-section.

ELECTROLYTE: aqueous solution of $Na_3SbS_3$.

ANODES: lead plates.

CATHODES: iron plates and the walls of the tanks.

CURRENT DENSITY: 10 to 15 amperes per sq.ft. at the start; E.M.F., 2 volts; later 4 to 5 amperes per sq.ft.

By this process not only is the solvent recovered in a commercial form, but also the sulphur of the ore.

Inasmuch as it is difficult to prepare antimony solutions perfectly free from Fe, there is always more or less FeS formed on the walls of the electrolyzing tank, and upon the surface of the cathode, and this falls to the bottom with the metal that drops off; hence this process does not yield directly a perfectly pure metal, but rather one that contains Fe.

# (C) Antimonial Lead, Hard Lead

In the purification of lead bullion (cf. Silver, Desilverization, and Lead) by an oxidizing fusion, one of the first oxidation products withdrawn is usually a mixture of PbO and $Pb_3(SbO_4)_2$. We have already seen in the above cited sections that this product is next subjected to a liquation during which a reducing flame

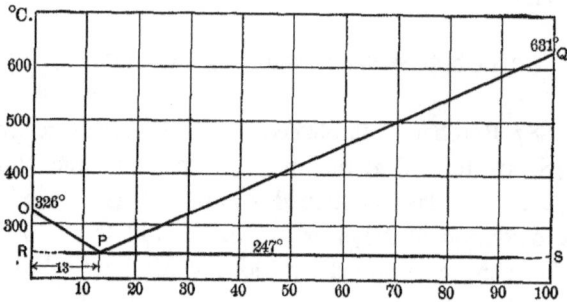

FIG. 161.

is maintained and small quantities of charcoal powder are spread over the charge. By this treatment it is aimed to cause the separation of mechanically enclosed Pb granules; but by means of the reducing flame and the charcoal powder a part of the PbO is reduced at the same time and the residue enriched in Sb. The further working up of the antimony skimmings is carried out, as was explained, by a

Reducing Fusion in small shaft furnaces, such as serve for the smelting of lead ores. In smelters in which but little of this by-product is obtained, low blast furnaces are also used for working up the litharge when convenient.

As a reduction product an Sb-Pb alloy is obtained with 14 to 20% Sb. The old assumption that hard lead is a chemical compound, $Pb_3Sb_2$, is untrue. Pb and Sb, as is evident from Fig. 161, are only soluble in one another as solids, and form a eutectic containing 13% Sb, which melts at 247° C.

# (D) Antimony Refining

The chief impurities of crude antimony are Fe (7 to 10%) S (up to 1%) and, as a rule, only small amounts of other metals. The refining of the metal, therefore, consists mainly in the removal of Fe, but during the process the other impurities disappear for the most part. Pb is very difficult to remove from Sb; ores showing a high percentage of Pb must be converted by separate smelting operations into an Sb-Pb alloy (hard lead) and crude Sb. The removal of the Fe and S is effected by the following

**Operations:**

1. **Sulphurizing Fusion** of the crude metal with $Sb_2S_3$ (8 to 12%) and a little common salt, NaCl (4 to 5%) in a crucible or reverberatory furnace of the same construction as described under Bismuth (cf. Figs. 130 to 137, pp. 112 and 114).

   By means of this fusion the metal is brought to a purity of 98 to 99% Sb with a low content of Fe and S. These last impurities are removed by the so-called

2. **Starring.** The metal, together with a strongly basic flux, composed of about 3 parts soda ash, or potash, and 2 parts $Sb_2S_3$, is again melted in either a crucible or reverberatory furnace. In the above case the $Sb_2S_3$ was used as an agent for the removal of Fe, but here, in the presence of alkali, it also had a tendency to form sulphantimonate and acts, therefore, as a desulphurizing agent as well. During this smelting it is of course necessary to take particular pains to avoid having the metal come in contact with iron; the iron instruments used for stirring and for ladling are, for this reason, always given a coat of enamel or whitewash, as far as there is any danger of their coming in contact with the metal. If, in the ladling or pouring out of the purified metal

from the vessel into the whitewashed iron molds, care is taken to scoop up and pour out enough slag to form a protective cover, then the crystallization of the Sb, which is associated with the cooling of the molten metal, will start at the walls of the mold. The crystals grow beneath the badly conducting slag layer and the last portions of the liquid collect at the surface, because they are there most protected from cooling; this liquid will be constantly drawn back by the crystalline framework which diminishes in surface area during the further cooling. Thereby the crystal dendrites that first form in it are pushed forcibly through the surface of the melt and " stars " are formed there, which are regarded by buyers as a sign of purity. This phenomenon has given the name of " starring " to the entire operation.

**Electrolytic Production of Pure Antimony.** This has been attempted under different working conditions, but up to the present time such processes have not been adopted on a large scale. Among such experiments may be mentioned the

**Electrolysis of Crude Antimony** with $SbCl_3.KCl$ solution as electrolyte (Sanderson) for working up auriferous antimony.

**Electrolysis of an $SbCl_3.FeCl_2$ Solution,** using insoluble anodes, with the idea that $FeCl_2$ will be converted into $FeCl_3$ at the anode, and that the latter will serve as solvent for the lixiviation of $Sb_2S_3$ ores.

ELECTROLYSIS: $6\ FeCl_2 + 2SbCl_3 = Sb_2 + 6FeCl_3.$

LIXIVIATION: $6\ FeCl_3 + Sb_2S_3 = 6FeCl_2 + 2SbCl_3 + 3S.$

The lixiviation, as well as the electrolysis, of such chloride solutions involve difficulties and high cost in providing sufficiently durable apparatus and in replacing the cathode metal upon which the $FeCl_3$ has a solvent action.

**Properties of Antimony:**

SPECIFIC GRAVITY: 6.7.

MECHANICAL PROPERTIES: brittle.

COLOR: white with high metallic luster.

STRUCTURE: pure antimony shows a large-leafed, crys-

talline structure, the surface of the so-called " antimony
button " showing star-shaped outlines (cf. Figs. 162,
163, 164).

FIG. 162.—Surface (×15).

FIG. 163.—Internal Grain Structure.

MELTING-POINT: 631° C. (1168° F.), readily volatile at
higher temperatures.

THERMAL AND ELECTRICAL CONDUCTIVITY: slight, the latter about 0.035 that of Ag. In the thermo-electric potential series, it is the most negative metal used in the construction of thermo-elements.

ALLOYS: particularly with the metals Pb, Bi, Sn, Cu, Ni, Fe, Cu; with the last four metals chemical compounds are also formed.

CHEMICAL BEHAVIOR. At ordinary temperatures it resists well the action of O and $H_2O$. Above its melting-point it oxidizes readily and burns to $Sb_2O_3$ and $Sb_2O_5$, or to $Sb_2O_4$, which may be regarded as containing half

FIG. 164.—Internal Dendritic Structure ($\times 40$).

a molecule of each of the other oxides. $Sb_2O_3$ shows both acid and basic properties, but $Sb_2O_5$ has an acid character. Sb combines directly with Cl to form $SbCl_5$ or $SbCl_3$. HCl dissolves Sb slowly. $H_2SO_4$ oxidizes Sb to $Sb_2O_3$ and an excess of this acid forms a sulphate, which hydrolyzes readily. $HNO_3$ oxidizes Sb to $Sb_2O_5$ Sb shows a great affinity for S; the sulphides $Sb_2S_3$ and

$Sb_2S_5$ (the latter is unstable in the free state) have an acid character and unite with alkali- and alkaline-earth sulphides to form sulphantimonites and sulphantimonates, which are readily soluble in water. In the fused state the affinity for S is so great in the presence of alkali monosulphides that the sulphantimonites, e.g., $Na_3SbS_3$, will withdraw S from even $Sb_2S_3$, with separation of Sb, to form $Na_3SbS_4$, which represents the pentavalent condition of antimony.

# NICKEL

## Sources

**Natural Sources:**

Native in meteoric iron, up to 20%.

SULPHIDE ORES.

MILLERITE: NiS (rare).
PENTLANDITE: (FeNi)S to NiS.2FeS.
HORBACHITE: (FeNi)$_2$S$_3$.
LINNÆITE: (NiCo)$_3$S$_4$.
GERSDORFFITE: NiAsS.
ULLMANNITE: NiSbS.

} usually with other sulphides, pyrite and pyrrhotite.

ARSENIDES AND ANTIMONIDES.

NICCOLITE: NiAs.
CHLOANTHITE: NiAs$_2$.
BREITHAUPTHITE: NiSb.

SALTS.

MORENOSITE: NiSO$_4$ + 7H$_2$O.
ANNABERGITE: Ni$_3$(AsO$_4$)$_2$ + 8H$_2$O.
ZARATITE: a basic carbonate.
GARNIERITE: a hydrosilicate of nickel and magnesium.

**Other Sources:**

Nickeliferous rocks, speisses and slags.

Metal clippings, metal waste, old metals.

## (A) Concentration Methods

Nickel in its ores is often associated with large quantities of gangue in such variety of ways that a direct smelting of the ore, wherever smelting comes into consideration at all, is out of the question.

160

As in the case of copper ores, and according to the same principles (cf. pp. 60 to 90), it is possible to carry out a concentrating process consisting of roasting and smelting operations, but in some cases it is more advantageous to employ a concentrating procedure which consists of roasting, lixiviation and precipitation.

Of the numerous proposals for concentrating nickel ores, the choice is determined, in the first place, by the presence or absence of copper, or by whether it is desired to effect a separation of nickel from copper; and, in the second place, by the question whether it is worth while to attempt a separation from cobalt. The removal of the other metals which are likely to be present in nickel ores offers no special difficulties and in no way influences the work of concentrating the nickel.

**Concentration without Separation of Copper.** By this work all copper-free ores can be handled, but if copper is present it is useful only when it is desired to prepare by smelting an alloy of copper and nickel rather than pure nickel. Next to nickel, copper has the greatest affinity for sulphur. For this reason the nickel ore can be concentrated by roasting and smelting to a matte just as was described under Copper. In the case of nickel, however, there is another possibility for concentrating, based upon the fact that nickel has the greatest affinity for arsenic of all the different metals that come into consideration here. Of course this method of smelting, the so-called *speiss smelting* (the arsenides of Ni, Co and Fe form what is called *speiss*), is applicable only when the ores, or metallurgical products at hand contain arsenic; in the case of arsenic-free ores the aim must always be to form a matte.

The mode of operating and the apparatus are almost exactly the same as in the case of copper concentration, namely

**1. Oxidizing Roast,** employed with sulphide ores for the same reasons as in the case of sulphide copper ores, and with arsenide ores in order to reduce the As content as far as 1 As to 2 Ni, but not any farther. A further reduction in the amount of As would lead to

Ni losses in the slag during the succeeding work; in smelting for crude matte or crude speiss, however, the metal content of the resulting slag must be sufficiently low to permit discarding of the slag.

Apparatus for carrying out the roast, cf., Copper.

PRODUCTS: $SO_2$ (eventually to be worked into $H_2SO_4$); $As_2O_3$ to be worked into white arsenic by condensation and resublimation; roasted ore.

2. **Formation of First Matte or First Speiss.** The roasted ore from the above treatment, or oxidized ore with the addition of sulphide (pyrites, tank waste, sulphates) together with rich Ni slags from the following operations and other waste products, are placed in a shaft or reverberatory furnace and subjected to a reducing smelt.

PRODUCTS: a slag low in metal content which can be discarded unless there is demand for it as building material, etc. A crude matte containing up to 40% Ni (perhaps Ni and Cu) and when low in Ni, up to 40% Fe. In the case of arsenical raw materials, a speiss is obtained with from 40 to 50% Ni.

3. **Oxidizing Roast of the Crude Matte, or Crude Speiss:** the purpose of this operation is the same as under 1.

APPARATUS: reverberatory furnace.

PRODUCTS: $SO_2$, $As_2O_3$ as under 1, roasted matte or speiss.

4. **Concentration of the Matte or Speiss** is carried out with acid fluxes in shaft or reverberatory furnaces as in the copper industry.

PRODUCTS: nickeliferous slag which must be again added to the charge in the smelting of crude matte, and a matte richer in Ni which still contains considerable Fe; this must be subjected to

5. **Blowing in Converters,** by which means a product containing 76% Ni is obtained.

The speiss from concentration is not refined, as a rule, but goes to the metal extraction processes with a Ni content of 65 to 70% Ni.

As regards the constitution of nickel matte, most metallurgical text-books give incorrect statements. The true constitution was established by Bornemann in 1907. The sulphide NiS is not capable of existence under the conditions prevailing in the smelting of matte and is never to be found even in the relatively rich mattes. According to the concentration of Ni, Fe, and S in the matte, there may be present combinations of $Ni_2S$ with FeS, of $Ni_3S_2$ with FeS, free $Ni_3S_2$, and when still less S is present there may be present a solution of Ni in $Ni_3S_2$.

Bornemann obtained the following combinations: $(FeS)_2Ni_2S$, which crystallizes from molten matte at 686° C. but is stable only at high temperatures and changes at 575° into $(FeS)_3(Ni_2S)_2$ with loss of FeS. At lower temperatures FeS is again taken up to form the compound $(FeS)_4Ni_2S$.

The poorer the matte is in FeS, the more of the compound $Ni_3S_2$ appears, as far as the concentration of the sulphur permits. In mattes that are free from FeS there are found, according to the concentration of the sulphur, $Ni_3S_2$, $Ni_2S$, and free Ni, the last two being held in solution by $Ni_3S_2$. Every experiment, in which it was attempted to form the compound NiS, so often mentioned as a constituent of nickel matte, was unsuccessful. Under the conditions prevailing in matte formation, the compound $Ni_3S_2$ represents the most sulphur that can be made to combine with nickel.

Furthermore, most statements in metallurgical literature previous to 1907 concerning the constitution of nickel speiss are likewise erroneous. Friedrich has been able to identify with certainty only two compounds between Ni and As, namely, $Ni_5As_2$ and NiAs; he believes, however, that the existence of a compound $Ni_3As_2$ is probable. The latter is formed, judging from the behavior of the melts examined, from mixed crystals of $Ni_5As_2$ with As, and the compound NiAs is often formed only after the complete solidification, under strong supercooling, of a eutectic corresponding to this composition. By inoculation (throwing in a crystal) it is possible to prevent the supercooling and bring about the solidification of the eutectic and the carrying out of the reaction according to which $Ni_3As_2$ is formed at a temperature lying above the eutectic point.

**Concentration of Nickel and Elimination of Copper by the Formation of Matte and Speiss.** If ores, metallurgical products, or metallic scraps are at hand which contain Ni, Fe, Cu, S, and As, a separation of the Ni and Cu can be effected, although not very sharply, if the roasting and smelting work is conducted in such a way that the material as it comes to the smelter contains enough sulphur to form a copper matte and enough As to form a nickel speiss. On account of the great affinity of Cu for S and of Ni for As, the greater part of the Cu will then be in the matte and the greater part of the Ni will be in the speiss. Although the matte and the speiss dissolve in one another to a slight extent, still for the most part they are obtained from the smelting in a separate state. The copper matte is, in other words, not free from Ni and the Ni speiss is not free from Cu, so that the work must be repeated with both of these products. The methods used in carrying out this work do not differ essentially from those described above for producing matte and speiss.

**Concentration of Nickel and Elimination of Copper by Smelting to a Matte.** This is employed in working up sulphide ores carrying Ni and Cu (pyrrhotite) of which large deposits occur in the Sudbury district, Ontario, Canada, and in Southern Norway. This ore also occurs in Southern Germany, but the deposits there have never been worked to any extent. For the last forty years a process has been used in some smelters which is quite complicated and consists of the following operations:

1. **Oxidizing Roast of the Ore.**
2. **Smelting for First Matte.** } as under 1 and 2, pp. 161–2.
3. **Matte Smelting** with the addition of $Na_2SO_4$ and charcoal.

APPARATUS: shaft furnace usually constructed with water jackets, cf. Johnson furnace, p. 80, Figs. 105 and 106.

PRODUCTS: Besides a waste slag, there is a Cu-Fe-Na matte and a specifically heavier Ni-Fe matte. After being tapped into slag pots, the two mattes separate into layers. The Cu-Fe-Na matte after solidifying forms the so-called " tops," and can be broken away from the other matte which constitutes the " bottoms." The tops resulting from this treatment are not yet poor enough in Ni to be worked for Cu, and likewise the bottoms contain too much Cu to be worked directly for Ni. It is necessary, therefore, to treat both of these products again.

4. **Smelting of the Tops.** The copper matte which contains sodium is first allowed to weather somewhat, the desired action being accelerated by wetting with the leachings from No. 7 (see below). Then, with the addition of some raw matte (from 2) it is subjected to a reducing smelt in a shaft furnace.

PRODUCTS: waste slag; tops consisting of a matte rich in Cu but poor in Ni, and a specifically heavier matte which has a Ni content corresponding to about that of the former bottoms with which it is smelted again.

5. **Smelting of the Bottoms.** The heavy mattes, obtained by operations 3 and 4, are smelted with sodium sulphate and charcoal in exactly the same way as in the third operation.

PRODUCTS: Besides a waste slag, a specifically light matte which corresponds in Cu content to the tops of operation 3, as well as to those of operation 4; with both of these it is again smelted. The specifically heavier matte forms a concentrated bottom which is rich in Ni and poor in Cu.

6. The concentrated tops are **leached with water.**

PRODUCTS: A solution of sodium sulphide and oxidation products; on evaporating to dryness, a mixture of $Na_2S$ and other Na salts is obtained which can be used again as a flux in the third operation. The leached residue consists chiefly of $Cu_2S$ together with practically all the

Au and Ag originally present; it is worked up in the usual way for Cu and the precious metals.

7. The concentrated bottoms are subjected to a **chloridizing roast** carried out in such a way that it yields, after treatment with water, the following

PRODUCTS: A solution containing the chlorides of the platinum metals and copper; these metals are recovered by precipitation in the usual manner. The leached residue contains a mixture of NiO with a little $SiO_2$, S, Cu, Fe and Pt. For the method for working this up, see Extraction.

**Concentration by Leaching and Precipitation,** if necessary, with the aid of roasting. Directly applicable to ores, intermediate products and by-products which contain the Ni in the form of easily soluble salts, or free oxides and sulphides. After a preliminary roasting it is particularly suitable for sulphide and arsenide ores and intermediate products which also contain besides nickel other metals worth recovering (Cu, Co and precious metals). Contrary to the many statements in metallurgical literature, it is to be emphasized that silicates are absolutely unsuited for such work.

**An Old Process of Roasting, Lixiviation, and Precipitation** was formerly used universally for working up mattes and speiss which contained Co, Cu and other metals besides Ni. It consists of the following separate operations:

1. DEAD ROASTING to Oxides.

2. SOLUTION OF THE ROASTED RESIDUE in acids (HCl or $H_2SO_4$).

3. PRECIPITATION by $H_2S$ or by a soluble sulphide ($Na_2S$, etc.).

PRODUCTS: Sulphides of Cu, Pb, Bi, Sb, As. These sulphides are utilized in the works according to the predominant metal.

4. OXIDATION of the residual solution with chloride of lime and neutralization with milk of lime. First of all, the iron is precipitated as $Fe(OH)_3$. After the greater part of the iron has come down, the solution is filtered and the

addition of chloride of lime and milk of lime continued. The cobalt is thus precipitated as hydrated $Co_2O_3$.

5. PRECIPITATION with Milk of Lime or Soda.

PRODUCT: $Ni(OH)_2$ or $NiCO_3$.

**Herrenschmidt's Process** for working up ores carrying Co and Ni. This process is used on Herrenschmidt's own property for *asbolite* (18% $Mn_2O_3$, 3% $CoO$, 1.25% $NiO$, 30% $Fe_2O_3$, 8% $SiO_2$, 5% $Al_2O_3$, 1% $CaO$, 1% $MgO$). It consists of the following operations:

1. OXIDIZING ROAST.

2. TREATMENT of the roasted residue with ferrous sulphate solution and introduction of steam and air.

PRODUCTS: $CoSO_4$ and $NiSO_4$ in solution.

3. PRECIPITATION of the solution, which is kept weakly acid, with $Na_2S$.

PRODUCTS: a small amount of CoS and NiS.

4. THE SULPHIDES are filtered off, dried and given a sulphatizing roast.

5. THE ROASTED PRODUCT is dissolved in water, and after treatment with $CaCl_2$ solution, $CoCl_2$ and $NiCl_2$ remain in solution while $CaSO_4$ is precipitated.

6. THE SOLUTION obtained in 5 is divided according to the ratio of Co to Ni. One part of this solution is completely precipitated with $Ca(OH)_2$ solution. The precipitate is converted into $Co_2O_3$ and $Ni_2O_3$ by means of air and chlorine. This precipitate is mixed with the remainder of the original solution so that the $Ni_2O_3$ reacts with the $CoCl_2$ in solution to form $Co_2O_3$ and soluble $NiCl_2$.

7. FILTRATION: the filtered precipitate when dried is $Co_2O_3$ which can be sold as such.

8. THE FILTRATE, containing $NiCl_2$ in solution, is precipitated by $Ca(OH)_2$.

9. FILTRATION: the filtered precipitate of $Ni(OH)_2$ is dried and ignited, whereby it is converted into NiO. This is worked up as described below under B. The liquors containing $CaCl_2$ can be used again in Operation 5.

**Borchers and Warlimont's Process** accomplishes in a simple manner a working up of Ni ores that carry Co and Cu. It consists of the following operations:

1. SULPHATIZING ROAST at a temperature of about 500° C., in rotating iron drums, into which the air necessary for roasting is led. The drums are heated externally.

PRODUCTS: $SO_2$, which can be utilized in the usual way, and a roasted residue in which the Co and Cu are present almost entirely as sulphates, the Ni partly as $Ni_3S_2$, partly as NiO and only to a slight extent as $NiSO_4$, while the Fe is converted for the most part into $Fe_2O_3$, although some $FeSO_4$ and $Fe(SO_4)_3$ remain.

2. EXPOSURE of the roasted product to the air for several days, meanwhile working it about and wetting it. Hereby the $FeSO_4$ that is present converts any unchanged $Cu_2S$ into $CuSO_4$.

3. LEACHING with slightly acid water.

PRODUCTS: a solution of $CuSO_4$ and $CoSO_4$, which is only slightly contaminated with $NiSO_4$ and $FeSO_4$, and a residue which contains most of the Ni together with the original gangue of the ore and the $Fe_2O_3$ that has been formed.

4. PRECIPITATION with iron.

PRODUCTS: cement copper, concerning the working up of which see Copper, and a weakly acid solution containing all the Co of the ore and a small amount of Ni, both in the form of sulphates.

5. PRECIPITATION of the solution obtained in 4, using CaS as precipitant; the latter is a by-product in the smelting of concentrated matte with CaO + C.

PRODUCTS: a mixture of CoS with a little $Ni_3S_2$ which is dried and used in the cobalt industry. The liquor obtained here may be evaporated to dryness and the residue used as a flux in operation 1.

6. **Matting** of the leached residue obtained in Operation 3 adding to the charge, when necessary, substances con-

taining S and free from Cu. At the same time other
nickeliferous slags and by-products obtained in subsequent operations may be added to the charge.

PRODUCTS: waste slag and first nickel matte which is
worked up further as described under A on page 162. As
regards the method of obtaining metallic nickel see the
following descriptions of Extraction and Refining.

# (B) Extraction

The choice of a process for working up the products obtained
by concentration work is determined by the character of the
material (whether a concentrated matte, a concentrated speiss,
or an oxide), the character of the finished product (whether pure
nickel, or a nickel alloy), and finally, by the availability of a cheap
fuel or other source of energy (water-power).

The Roast-Reduction Process has in the past been most used
because the concentration work, as described above, has
yielded chiefly a concentrated matte or speiss. This process
comprises the following operations:

1. **Dead Roast.** Matte or speiss is roasted with great care
in a reverberatory furnace until the last traces of S or
As have been oxidized and removed. Sometimes it is
necessary to add a little $Na_2CO_3$ or $NaNO_3$ (or a mixture
of both salts) in order to remove the last traces of As.

APPARATUS: reverberatory furnaces.

PRODUCTS: $SO_2$, $As_2O_3$, which, however, on account of the
low concentration in which they are present in the gases,
cannot be utilized; hence, wherever this method is employed
its effect is very injurious to vegetation in the adjacent
country. The roasted residue consists almost entirely
of NiO. In case one or both of the above-mentioned
Na salts were used to effect the removal of the As, the
residue also contains $Na_3AsO_4$, which can be extracted
with water.

2. **Reduction.** For this work, there is the residue obtained

under 1 as well as oxidized products obtained in concentrating work as described under A. The reduction may consist either of a reducing roast or of a reducing smelt. The reduction by means of a reducing roast is effected by taking the NiO obtained in previous operations, stirring it up with flour paste to a plastic mass, rolling it into plates, and cutting these lengthwise and crosswise into small cube-shaped briquettes, which, after being dried in retorts or crucibles, are embedded in charcoal powder and heated within the same vessels to the reduction temperature of NiO, usually using producer gas as fuel. It is not intended hereby to effect a complete fusion, but merely to weld together the reduced Ni; the cubes, naturally, are smaller in size than the cubes of NiO that were introduced into the reduction furnace. The " cube nickel " thus obtained is preferred in this form by buyers for many purposes. It is, for example, suitable for purposes of alloying, but it is not suitable for working into nickel wares on account of the presence of carbides as well as some NiO, which impurities cause brittleness. A direct reducing smelt of nickelous oxide, which may take place in a crucible furnace or in a shaft furnace, is usually employed when the oxidized concentration product still contains metals that should either be eliminated by further separation processes (cf. Nickel Refining), or when it contains large amounts of other oxides (CuO), which it is desired to reduce at the same time so that a nickel alloy will be formed.

**Roast-Reaction Smelting** in reverberatory furnaces or in converters does not give rise to metal, as is the case when working with copper products. The reason for this was found, in the author's laboratory, to be as follows: The reaction

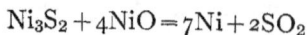

$$Ni_3S_2 + 4NiO = 7Ni + 2SO_2$$

takes place, to be sure, at temperatures above 1400° C., but only very slowly even when the heat is raised above 1600° C., and with high oxygen concentration in the blast. On the the other hand the reaction

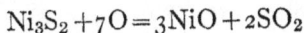

$$Ni_3S_2 + 7O = 3NiO + 2SO_2$$

takes place rapidly even at relatively low temperatures. Moreover, metallic Ni alloys very readily with its sulphides and with NiO.

Reaction smelting, therefore, yields under the most favorable conditions only an alloy consisting of much NiO with little Ni; it can never lead to metallic nickel under any of the methods of working which have been tried up to the present time.

**Desulphurizing Fusion of Concentrated Matte.** Borchers and Lehmer's process consists of smelting the concentrated matte with lime and carbon, preferably in an electric furnace. If the matte was entirely free from other metals, then this smelting can be carried out so that a fused, pure nickel will be obtained directly. Since, however, the sulphide nickel ores usually contain precious metals (Au and members of the Pt group), the product is an alloy of Ni with these metals, from which pure nickel is obtained by electrolysis (see Nickel Refining). The slag of CaS can be used either as a sulphide flux in the smelting of matte, or, as in the above-described process of Borchers and Warlimont, as a precipitant for Co and Ni solutions.

# (C) Nickel Refining

As a raw material for the preparation of pure nickel, the crude Ni prepared as described under B is usually taken. It is possible, however, to make use of concentrated matte, or even certain ores, as starting material.

**Pure Nickel from Crude Nickel.** The crude nickel obtained by a reducing roast, or the nickel obtained by a reducing smelt of nickelous oxide, contains, as has already been mentioned, C and NiO so that it is not suitable to be used in the manufacture of nickel wares (foil, wire, etc.).

**Refining Fusion.** The removal of the C offers no difficulties, as it takes place when the crude metal is melted; but the last traces of NiO cannot be removed by means of carbon. This reduction of the NiO was first accomplished by the method of Fleitmann, who fused with Mg, which is an extremely energetic reducing agent; then Basse and Selve accomplished the same end by mixing in as much as $3\%$ of $MnO_2$ before carrying out the reducing roast. The $MnO_2$ is reduced, together with the NiO, and serves to remove the NiO when the cube nickel is heated in carbon-free crucibles; naturally it prevents the formation of NiO by atmospheric oxygen when the metal is melted.

**Electrolysis** is employed if the nickel contains precious metals. Since the solution tension of Ni is on the other side of H, the concentration of H ions must be kept as low as possible, contrary to the practice in the electrolytic refining of Cu. This is accomplished by limiting the amount of free acids having high dissociation constants (HCl, $H_2SO_4$) or by using acids with low dissociation constants ($H_3BO_3$, $H_3PO_4$, and organic acids). Unfavorable changes in concentration are avoided by moving the electrodes and heating the electrolyte. When these requirements are fulfilled, the electrolytic deposition of Ni offers no difficulties worth mentioning. The conditions for carrying out the electrolysis are as follows:

ELECTROLYZING TANKS: made of wood with lead lining.

ANODES: Cast plates of impure nickel, containing precious metals with S up to $3\%$.

CATHODES: Pure Ni.

ELECTROLYTE: $NiSO_4$ or $NiCl_2$ solution, the former being

preferred.   Concentration 4 to 14 oz. Ni to the gallon
not under 0.01% or over 0.25% of free acid.

TEMPERATURE: 50° to 90° C. (122° to 194°F.)

CURRENT DENSITY: 5 to 28 amperes per sq.ft.

POTENTIAL: 1 to 1.3 volts with 15 or 20 amperes per sq.ft.

REACTIONS DURING THE ELECTROLYSIS: under the above
conditions it is possible, even when the crude metal contains up to 0.5% Cu, to deposit a sufficiently pure Ni
(0.1% to 0.2% Cu).   It is important to have enough S
present not only to precipitate the Cu as $Cu_2S$, but also
to make the last traces of Fe form compounds of $Cu_2S$
with FeS.   Such compounds are attacked by the electrolysis less in proportion as the current density at the
anode is kept low, i.e., the more the anode surface is
increased in proportion to the size of cathode surface.   In
designing an electrolyzing plant for Ni, therefore, the anodes
should be made relatively large and the cathodes small.

**Pure Nickel from Concentrated Matte.**   The matte which
is obtained as end product in the concentration of sulphide
ores contains up to 76% Ni.   It then has 23% to 24% S, up to
0.4% Fe, 0.1% to 0.2% Cu, and small amounts of $SiO_2$ due
to the presence of a small amount of enclosed slag.   According to the above-cited researches of Bornemann, unlike the
concentrated matte of copper smelting, it does not contain
a pure sulphide of nickel, but rather a solution of Ni in $Ni_3S_2$.
This matte already has many metallic properties and is suited,
as has been demonstrated by Borchers and Günther, for a
**Direct Electrolytic Treatment.**   The conditions, as established
by Günther are partly the same as in the electrolysis of
crude Ni.   The points of difference may be summarized as
follows:

ANODES: concentrated matte, cast into plates.

CURRENT DENSITY: 23 to 26 amperes per sq.ft.

POTENTIAL: 3 volts.

REACTIONS DURING THE ELECTROLYSIS: Of the constituents
of the matte, it is chiefly Ni that passes into solution at

the anode, leaving behind free S, which remains adhering to the electrode as a coherent but porous crust until the matte has nearly disappeared, and thus influences the progress of the electrolysis scarcely at all. In this crust there are present about 80% free S and 20% of a mixture of Cu, Fe and Ni embedded in it. Whereas the original matte contained less than 0.2% Cu, the mixture of embedded sulphides will contain over 12% Cu with about 51% Ni and 3.5% Si+C. From this composition, and from the low Cu content of the Ni that is deposited on the cathode, it follows that Cu is combined with S in the crust and that the latter unites with the sulphides of Fe and Ni to remain, for the most part, adhering at the anode; this is confirmed by the results obtained in the electrolysis of crude Ni that contains S.

**Electric Smelting of Concentrated Nickel Matte** with lime and carbon directly yields, as already mentioned under B, pure Ni provided the matte contains no foreign metal other than S and Ni.

## Pure Nickel Directly from Ores.

**Extraction of Ni by Means of CO, Mond's Process.** This is based upon the property that Ni possesses of combining with CO to form a readily volatile compound called nickel carbonyl, $Ni(CO)_4$. The process is applicable only to ores and roasted products that contain the Ni in the form of free oxides. The operations are as follows:

1. REDUCING ROAST in retorts heated to 300° C. with gases containing H. The NiO is changed by this treatment into porous Ni. It is important to maintain a reducing temperature as low as possible so that a large surface of the reduced Ni will be obtained.

2. PASSING OF GASES CONTAINING CO at 100° C. and 15 atmospheres pressure over the charge. The temperature of 100° C. is chosen because it can be kept constant easily by means of steam. As regards the pressure, Dewar found that $Ni(CO)_4$ is stable at

50° under a pressure of　2 atmospheres.
100°　　"　　　"　　15　　"
180°　　"　　　"　　30　　"
250°　　"　　　"　　100　　"

3. Deposition of Ni from the Ni(CO)₄, which depends upon the ready reversibility of the reaction

$$Ni + 4CO \rightleftarrows Ni(CO)_4.$$

The vapors of $Ni(CO)_4$ escaping from the vessels under pressure can be decomposed into Ni and CO either directly, or, after a preliminary purification which consists in cooling and freeing from mechanical admixtures, by simply heating at 200° C. under ordinary atmospheric pressure.

The CO is collected and used again in the process.

## Properties of Nickel:

SPECIFIC GRAVITY: 9.

COLOR: light gray.

MECHANICAL PROPERTIES: possesses a high degree of ductility and tenacity so that it can be rolled into foil and drawn into wire.

STRUCTURE of Cast Ni: similar to that of ferrite (see iron); rolled Ni is very fine grained.

MELTING POINT: 1451° C. (2644° F.).

THERMAL AND ELECTRICAL CONDUCTIVITY: about 0.2 that of Ag.

ALLOYS: with most metals. Important alloys are coin metal, Ni and Cu; German silver, Ni, Cu and Zn; nickel steel, Fe and Ni. In the extraction of Ni the tendency to alloy with its own sulphides and arsenides, as well as with NiO, comes into consideration.

CHEMICAL BEHAVIOR: The metal resists oxidation fairly well at moderate temperatures. Waste in forging and rolling is slight. Solution tension in acids and salts is greater than H($+$0.223 volts toward H).  It dissolves, therefore, quite slowly in HCl and $H_2SO_4$, more rapidly in $HNO_3$, forming under normal conditions a bivalent Ni cation. In the molten condition the solution tension is greatest in arsenides, next in sulphides; in arsenides the solution tension is greater and in sulphides less than that of Cu.

# IRON

## Sources

**Natural Sources:**

NATIVE, as a constituent of the earth's crust it is of rare occurrence; large lumps of meteoric iron are, however, sometimes found (up to 55,000 lbs. in weight).

AS OXIDE AND HYDRATED OXIDE in the following ores:

Specular Iron Ore ⎫
Hematite ⎬ $Fe_2O_3$ with 70% Fe.
Red Iron Ore ⎭

Magnetite, or Magnetic Iron Ore: $FeO.Fe_2O_3$ with 72.4% Fe.

Brown Iron Ore ⎫
Limonite, Bog Ore ⎬ $Fe_2O_2(HO)_2$ to $Fe_2O(OH)_4$.
Brown Hematite ⎭

As SULPHIDE in

Pyrite ⎫
Marcasite ⎬ $FeS_2$.

Magnetic Pyrites, or pyrrhotite, FeS with some $FeS_2$, usually nickeliferous. Combined and mixed, as FeS with $FeS_2$, in numerous other sulphides.

The sulphides themselves do not come into consideration as ores of iron until after they have been utilized for the manufacture of $H_2SO_4$; they are then roasted to oxides so that they retain but very small amounts of sulphide. (Roasted or "burnt" pyrite, see below).

SALTS.

Spathic Iron Ore, or Siderite, $FeCO_3$ with 48.27% Fe, usually carrying some $MnCO_3$, $CaCO_3$ and clay.

Blue Iron Ore, or Vivianite, $Fe_3(PO_4)_2$, which seldom

177

occurs by itself, but accompanies other ores (e.g.,
limonite).

**Other Sources:**

METALS: Scrap from the mechanical working of iron and
old metal.

OXIDES: Roasted pyrite, the roasted residue from sulphide
ores. Hammer scale, the oxide crust formed on iron
heated to redness and detached in the mechanical working.

SALTS: Basic silicates and phosphates, slags from the refining
of iron.

# (A) Concentrating and other Preliminary Operations.

**For Concentrating Iron Ores** it is often desirable to employ:

**Magnetic Concentration** which is carried out largely when the
ore is magnetite. Ores carrying hematite are also concentrated by the magnet, after they have been converted into
magnetite by roasting.

**The Briquetting** of pulverulent ores and other raw material has
been carried out to a considerable extent in recent years. The
requirements of a good ore briquette for blast furnace work,
have never been satisfied.

**Chemical Preparation.**

**Dissociating Roast, usually combined with an oxidizing roast,**
is carried out to effect the

DEHYDRATION of hydrated ores (limonite).

OXIDATION of magnetite.

DECOMPOSITION AND OXIDATION of carbonates (spathic iron
ore).

WORKING up of pyrite for sulphuric acid.

The **Apparatus** used for this work consists of simple shaft
furnaces, the so-called *calcining furnaces* which usually work
with natural draft. They either take both ore and fuel, or
the ore alone, when they are fired externally with solid or
gaseous fuel. The discharge is usually through a saddle-

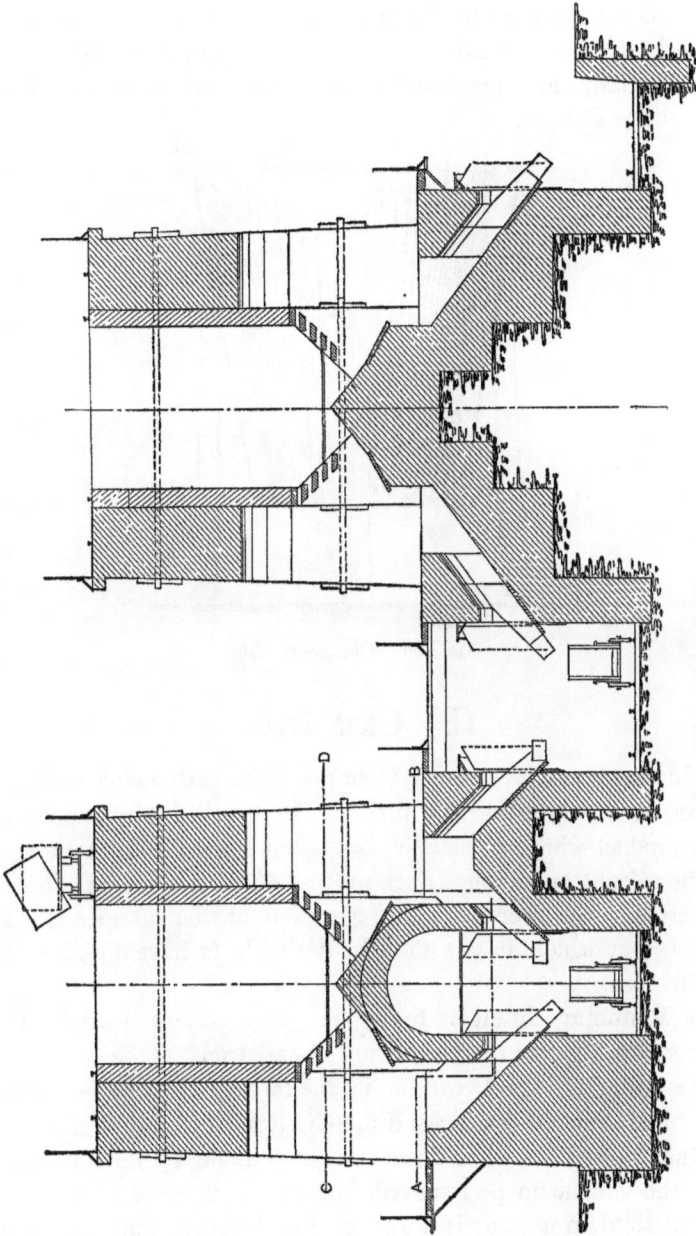

FIG. 165.—Roasting Kiln for Spathic Iron Ore. Scale 1 : 125

shaped bottom to facilitate the taking out of the solid charge through side openings. The apparatus that serves specially for the roasting of pyrite has been described under Copper.

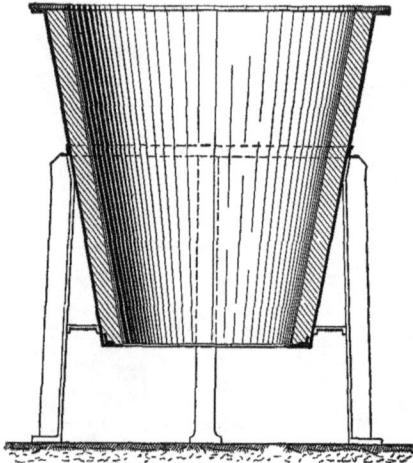

FIG. 166.—Old Siegen Kiln

# (B) Cast Iron

**Reducing Roast.** Various attempts have been made to make use of this operation and produce the so-called "iron sponge," a product which is metallic, but is not fused until it comes to the refining operation. Such processes have never been adopted permanently, even for the preparation of iron for special purposes in which it was thought desirable to have the iron in this form.

**The Reducing Smelt** is to-day the only process used in the iron industry for the production of cast iron.

**As Reducing Agent,** carbon in the form of charcoal is seldom used, but rather coke and the CO formed in the furnace.

**The Other Fluxes** are determined not alone by the nature of the gangue to be removed, but by the purpose which it is desired to accomplish, and by the demands that are to be placed upon the iron. Thus for slagging off the $SiO_2$ and clay which is frequently present in the ore, CaO in the form of $CaCO_3$ is added to the charge, and the amount added depends upon whether it is desired that the cast iron produced shall have little or much Si. To produce a cast iron rich in Si, it is necessary to form a more basic silicate and a slag richer in $Al_2O_3$ than when an iron low in Si is desired. For slagging off the S from ores containing some sulphide, the slag should be kept as basic as possible with CaO. The addition of an ore rich in Mn can also serve to advantage in this case.

**For Fuel,** coke is generally used and charcoal but seldom. In some few cases where coke is expensive and the ore deposits are in the vicinity of cheap water power, electricity is now to be considered as a possible source of heat. The blast requisite for the production of the desired amount of heat by the combustion of carbon is given a preliminary heating of 750° to 900° C. and compressed up to one atmosphere.

**Apparatus:**

BLAST FURNACES. These shaft furnaces are sometimes as much as 100 feet in height. The shaft rests upon pillars and is built of masonry with iron bands, or of an iron casing, and has an inside diameter of 26 ft. at the widest part, narrowing toward the throat, which has about two-thirds this width. The iron blast furnaces usually have internal crucibles. The hearth has an inside diameter of 13 ft. and a maximum height of $11\frac{1}{2}$ ft. It contains the tuyeres for the blast, the cinder notch, and the tap hole. One tuyere is reckoned for each sq. yd. of hearth surface so that with a diameter of 12 ft. there are 11 or 12 tuyeres. These tuyeres are hollow bronze or copper castings which can be cooled. The boshes rest upon the hearth and reach into the shaft to about two-fifths of the whole height of the furnace, reckoning from the bottom of the hearth, and they widen as they go upward until they reach the widest part of the furnace. The boshes and hearth are both cooled by spraying with water. Large furnaces use about 525 gallons of water per minute. Iron blast furnaces work with a closed throat, for the hot escaping gases are rich in CO and can be utilized as a source of heat and power. The "down-comer" is usually a wide

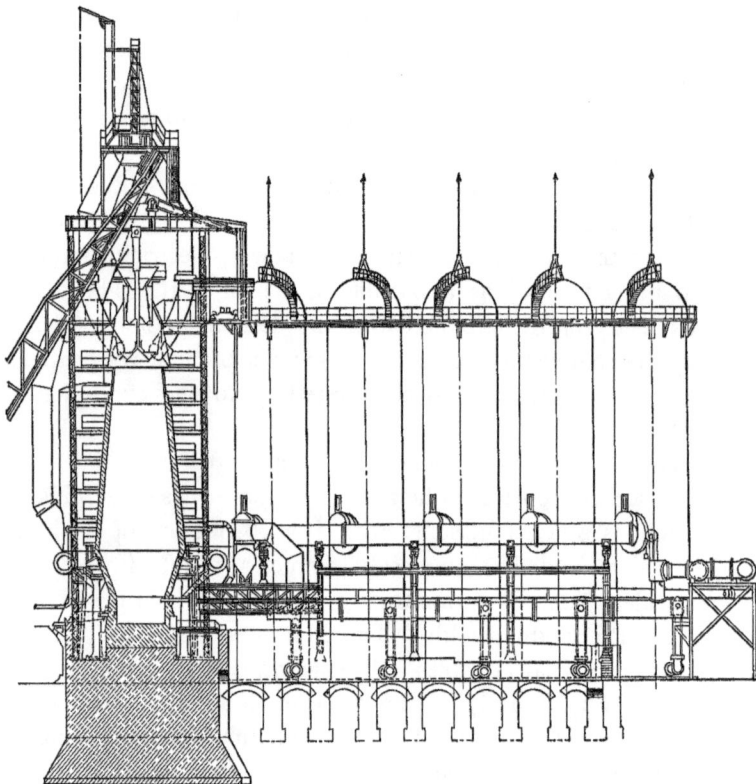

FIG. 167.—Blast Furnace with Stoves (Friedrich-Alfred Smelter) Rheinhausen.

central sheet-iron tube passing through the bell and hopper-feed in the throat. In some constructions, the gas is taken off laterally from the closed throat of the furnace.

HOT BLAST. To-day regenerator stoves of about the same dimensions as the blast furnaces usually form an

inseparable whole with the latter throughout all of
their operations. The regenerative system requires the
setting up of at least three stoves; to be on the safe
side it is customary to allow four stoves for each blast furnace. In the cylindrical stoves of Cowper's design, there

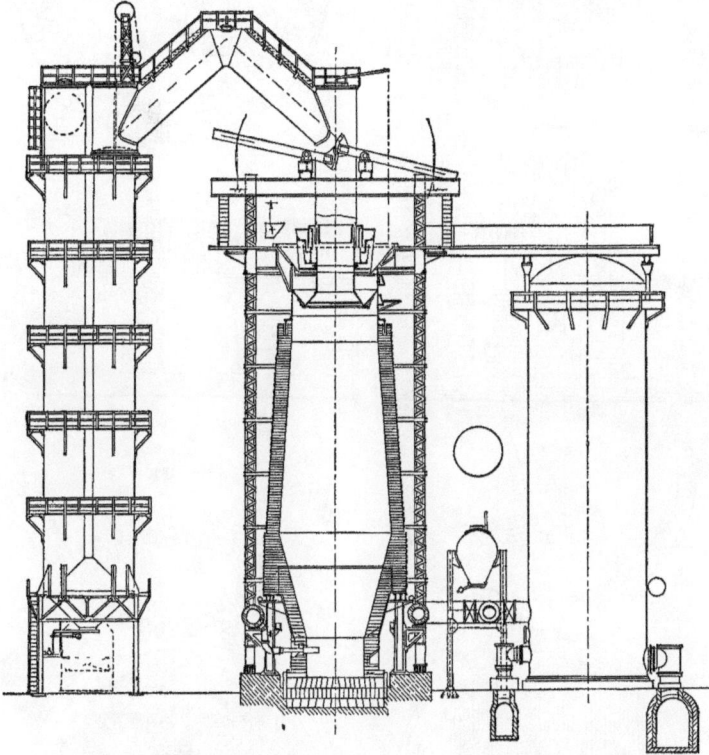

FIG. 168.—Blast Furnace with Dust Catcher and Blast Stove.

is found a shaft into which the hot gases from the throat of
the blast furnace enter at the bottom for the CO to be
burned by air which has been heated by passing through flues
in the walls of the shaft. The remaining space of the regenerator is loosely filled with special brick, the spaces between
which form numerous little channels. It forms the real

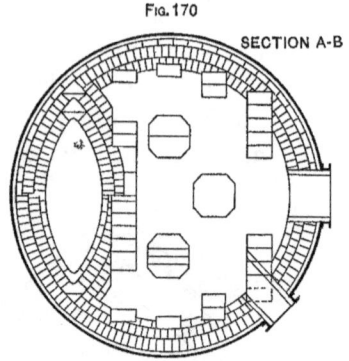

Fig. 170

SECTION A-B

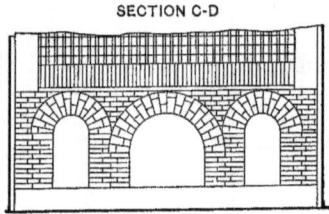

SECTION C-D

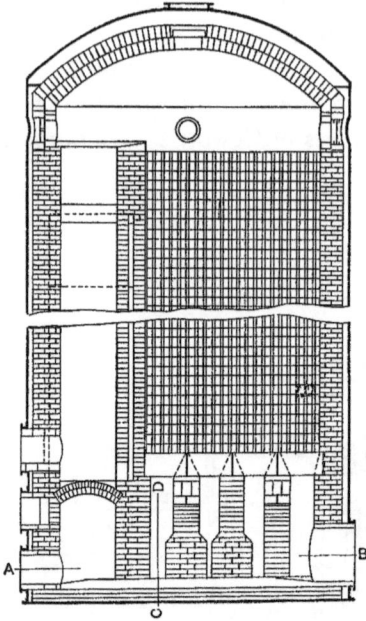

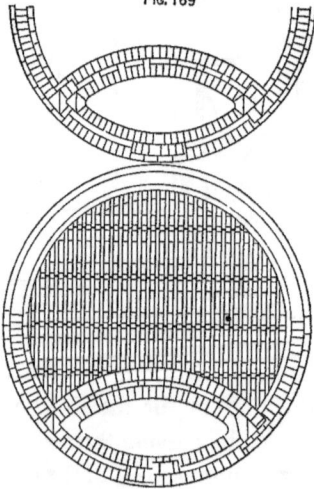

Fig. 169

Fig. 171

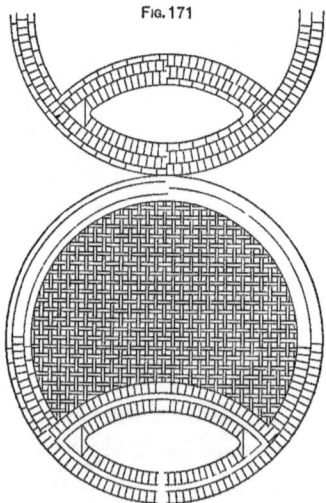

Fig. 172.

Fig. 173.

heat reservoir. The products of combustion are divided
into innumerable gas currents by these brick, and in this
part of the shaft they are carried downward and then led
away from the open space at the bottom. After the stove
has become hot enough, the gas and combustion air are
led into a neighboring regenerator, and now the air intended
for the blast furnace is passed through the hot chamber in
the opposite direction to that previously taken by the
gases from the blast furnace. This heated air collects
above the part of the regenerator that is filled with brick
and passes downward through the combustion chamber
and thence into the main pipe for the blast. Since this
air must be heated to between 750° and 900° C., the iron
pipes through which it passes are provided with a firebrick lining. Even the bustle pipe leading to the separate tuyeres of the blast furnace is lined or else placed
in a hollow tube with non-conducting layer of air inter-
vening.

REACTIONS IN THE BLAST FURNACE. The changes within
the more important temperature boundaries, which the
solid charge experiences as it passes from the throat of
the furnace downward, are the following: With furnaces
that are 75 to 100 ft. high between the throat and the
bottom of the hearth, the temperature at the top ranges
between 150° and 300° C., whereas in the smelting zone
the temperature may reach 1500°.

150° to 400° C.

Drying of the charge. Breaking down of hydrates. Beginning of reducing reactions.

400° to 1000° C.

According to the studies of Boudouard on the equilibrium
of the reaction

$$2CO \rightleftarrows CO_2 + C,$$

the speed of the reaction in the direction of the upper arrow increases from about 400° to 700°, but from there on the tendency for the reaction to take place in the opposite direction becomes more and more pronounced. Here the efficiency of CO as a reducing agent rapidly diminishes. As regards the concentration of the reaction gases, it has been found that the ratio of $CO:CO_2$ must not be less

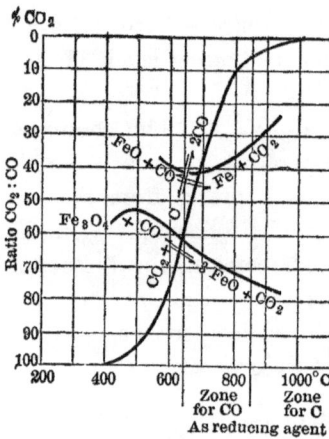

FIG. 174.

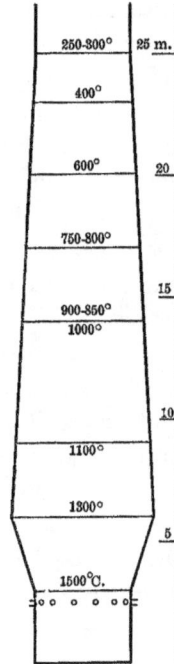

FIG. 175.

than 1:1 at the most favorable reaction temperature (700°). At such a concentration of $CO_2$, the reducing action of the CO stops almost entirely even at 700° C.

In this zone, therefore, the greater part of the oxides of iron, namely $Fe_2O_3$, $Fe_3O_4$, and FeO, are reduced by CO, but at about 800° there is a dissociation of the carbonates, $CaCO_3$, $FeCO_3$ and $MnCO_3$, into CaO, FeO,

$MnO$, and $CO_2$, and the effect of this is to greatly increase the concentration of the $CO_2$.

## From 1000° upward

the C begins to dissolve in solid Fe. Disregarding other influences (Si, Mn, etc.) it has been established that the solubility of C in Fe increases as the temperature rises (cf. also p. 190).

## From 1300° upward

solid C is practically the only reducing agent, particularly for $FeO$, $MnO$, $SiO_2$, phosphates, and silicates; FeS also reacts with CaO and C to form Fe and CaS.

The blast, as it passes through the furnace in the opposite direction, of course loses its O rapidly by reason of the combustion of the C. The more the blast has been heated before it comes in contact with the coke, the greater will be the formation of $CO(C + O_2 = CO_2$ and $CO_2 + C = 2CO)$. The CO, after it is formed, as the above equations show, will be changed repeatedly into $CO_2$ and back again to CO. The gas finally leaving the throat of the furnace has a composition of 22 to 28% CO, 16 to 8% of $CO_2$, 63 to 57% N, and a low per cent of hydrocarbons and free hydrogen.

With the present-day dimensions of blast furnaces, about 500 tons of pig iron are produced in a furnace during a day of 24 hours. The weight of a single charge of ore (ore and slagging fluxes) reaches 15 tons, for which 3 to 7 tons of fuel are required. The charge remains in the furnace for from 18 to 26 hours. Aside from the content of the ore and the quality of the other constituents of the charge, the capacity of the furnace is dependent particularly upon the nature of the pig iron that it is desired to produce. A furnace capable of producing 100 tons of white cast-iron, can yield at the most but 80 tons of gray iron, and still less of other kinds, e.g., metal running high in Mn or Si. Furthermore, in respect to the consumption of fuel, the production of white iron is the most favorable. For such iron, 100 to 110 tons of coke are required to produce 100 tons of the metal. The coke consumption rises with other qualities of iron. Concerning the nature and value of the

**Products of the Blast Furnace,** we may summarize them as follows:

1. PIG IRON, an alloy of Fe and compounds of Fe and Mn with C, Si, P, S, and O, from which on solidifying individual constituents, such as C in the form of graphite or amorphous carbon, carbides, silicides, etc., may separate out. The total C content is over 2% and usually under 5%.

The nature of the ore, the conditions of working, and the qualities required by the purchaser, have led to the production of quite a number of different commercial grades of pig iron, but the qualities, as in the case of most pure irons, are dependent largely upon the amount and manner of combination of the C present. The latter, on its part, is influenced by other constituents found in the iron, particularly Si and Mn; a short discussion of the relations between Fe and C will explain best the nature of the different products obtained in the iron industry.

If we start with a perfectly liquid melt, it is known from the interesting researches of Roberts-Austen, Wüst and Goerens, and others, that all the C can exist combined with Fe as $Fe_3C$ (which is called cementite) until the amount of C present reaches 6.66%. A higher C content scarcely comes into consideration in practical work; with a lower C content there may exist in the fusion a solution of this carbide in Fe. If the temperature falls, it may happen when the amount of $Fe_3C$ present is large that, just before the whole melt solidifies, a decomposition of the $Fe_3C$ into $3Fe+C$ takes place.

Directly after such decomposition free C is therefore embedded as pure graphite in pure Fe, and as the melting-point of the latter is higher than that of the solution (about 1500° C) it begins to solidify within the liquid melt and thereby envelopes the solid graphite. As long as the latter possesses this specifically heavy shell it remains uniformly distributed in the melt. When, therefore, the cooling does not take place too slowly, a pig iron is obtained which contains large crystals of graphite uniformly distributed through it. When, however, the temperature is kept high for a long time, the crust of Fe gradually becomes saturated again with C, becomes liquid and the graphite now rises to the surface and forms a scum called " kish." The melt, in which all this has taken place, finally solidifies on further cooling at 1130° C., with about 4.2% C chemically combined with Fe.

From molten iron with less than 4.2% C and more than 2% C, crystals first separate from the melt which are to be regarded as solid solutions of $Fe_3C$ in Fe (mixed crystals). The saturated solution, or in other words the mixed crystals richest in C, contain 2% of C, they correspond, therefore, to a solid solution of about $2 Fe_3C + 15Fe$.

The mother metal from which these crystals are deposited thus becomes enriched in $Fe_3C$ until the above-mentioned 4.2% C content is reached. Then the remaining metal solidifies at 1130° C.

At the concentration 4.2% C, lies a eutectic point between the melting points of the saturated mixed crystals, $2Fe_3C + 15Fe$ (2% Fe) and cementite, $Fe_3C$ (6.66% Fe), and this eutectic corresponds closely to the composition $Fe_3C + Fe_2$. As Wüst and Charpy both showed independently in 1905, this eutectic solidifies under all conditions, no matter what other substances may be present, at 1130° C.

Under the conditions corresponding to the line $aB$ (Fig. 176), there exists a solid mass of free mixed crystals embedded in eutectic; at the point $B$ there is only eutectic present; and at the line $BC$ graphite and cementite in eutectic.

In every cast iron with more than 2% C, there tends to take place after solidification a decomposition of $Fe_3C$ into $3Fe + C$ (graphite or amorphous carbon, temper carbon) provided the $Fe_3C$ has already crystallized from the solution or has had an opportunity to crystallize. Substances which favor the crystallization of this compound from the liquid or solid solution (e.g., Si, Al?) naturally favor as well the progress of the reaction $Fe_3C = 3Fe + C$; substances

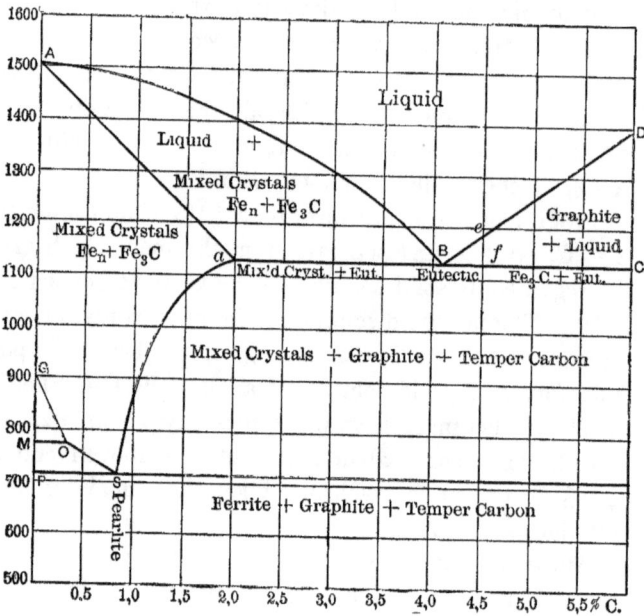

FIG 176.

which increase the solvent power of Fe for $Fe_3C$ (e.g., Mn) increase the stability of the $Fe_3C$ and hence tend to prevent the formation of free C. From this point of view the nature of the two principal kinds of cast iron appears in the following light:

WHITE IRON is a supercooled solution and is, therefore, to be regarded as representing a meta-stable system between $Fe_3C$ and Fe, in which the reaction $Fe_3C =$

$3Fe + C$ has not been allowed to take place. The chilled white iron consists of cementite and pearlite (see Fig. 178), also containing, when the cooling was very rapid, some mixed crystals of the concentration $Fe_3C : Fe_n$.

If such an iron is kept for a long time at about 1000° C. the compound $Fe_3C$ begins to crystallize from the solid solution and then the $Fe_3C$ decomposes rapidly, although the free C does not appear as graphite, but as the amorphous, readily-oxidizable, temper carbon (see Fig. 180, p. 195).

GRAY IRON represents the more stable system $Fe$—$Fe_3C$—$C$. It has had time, at the different temperatures and concentrations, to reach more or less completely a state of equilibrium, i.e., during the cooling $Fe_3C$ has had opportunity to crystallize out and dissociate into $3Fe$ and C and the latter is now to be found in the form of graphite, although it is not impossible that some temper carbon may have been formed in the vicinity of 1000°. (Fig. 179.) As has been pointed out, the separation of C is favored by the presence of Si, even when the C content is low, and to promote this separation from 1 to 2.5% Si, or even 5% Si in special cases, is often added to the iron.

Whereas gray iron is used for foundry purposes as well as for the production of wrought iron, white iron is used almost exclusively for the latter purpose. A distinction is further made between foundry iron and Bessemer iron (Si high, P low), open hearth iron, basic Bessemer iron (high in Mn and P), spiegeleisen or ferromanganese (rich in Mn, low in P), and charcoal iron (rich in C).

In the many constituents of the different kinds of pig iron, only a few of the particularly important solidification phenomena have been explained. The first accurate equilibrium diagrams between Fe and C were those of Roberts-Austen and of Roozeboom. Subsequently Heyn gave different diagrams for the meta-stable and stable systems, while Benedicks improved the original diagram of Roozeboom. Since Wüst and Charpy have determined the freezing-point of the system Fe—Fe₃C to be 1130° for all cases, and Wüst and Goerens have proved that graphite is always produced by the decomposition of Fe₃C, Goerens' diagram (Fig. 176) may be taken as correct for the stable and for the super-cooled system Fe—Fe₃C. To understand the terms used in metallography, the following definitions will serve:

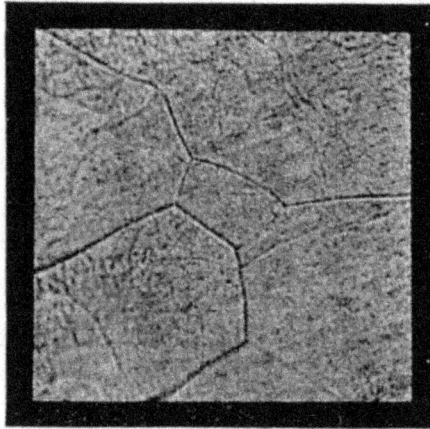

FIG. 177.—Ferrite (×500).

FERRITE, chemically pure iron: α-iron, which is magnetic and free from C, passes at 780° into β-iron, which is non-magnetic and also practically incapable of dissolving C. This second form of iron exists only between 780 and 800° C. Above 880° it changes to γ-iron, which is likewise non-magnetic, but is capable of dissolving C or Fe₃C (Fig. 177).

CEMENTITE is iron carbide, Fe₃C (Fig. 178).

AUSTENITE and MARTENSITE are solid solutions of Fe₃C in γ-iron. They exist as mixed crystals of varying concentrations. The saturation point of these solutions lies at 2% C corresponding to the composition 2Fe₃C +

15Fe.   A euuectic between these saturated mixed crystals and $Fe_3$ C lies at 4.1% C, corresponding to a composition $Fe_3C + 2Fe$.

TROOSTITE: a colloidal solution of $Fe_3C$ in Fe.

SORBITE: mixtures of Fe, $Fe_3C$, and solid solutions of $Fe_3C$ in Fe.

PEARLITE: the eutectic between ferrite (Fc) and cementite ($Fe_3C$). Its C content is 0.9%, corresponding to $Fe_3C + 2oFe$ (Figs. 178 to 180).

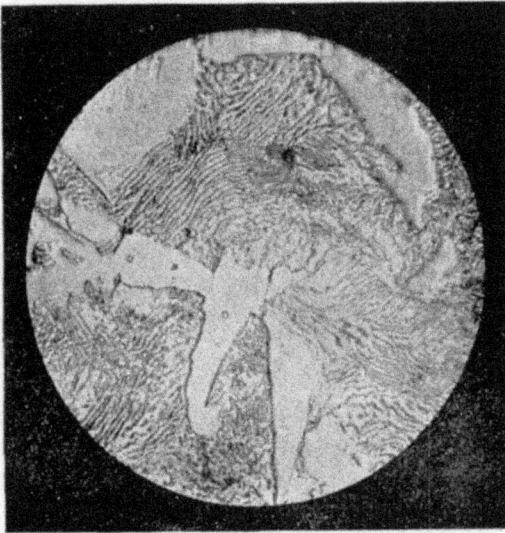

FIG. 178.—White Cast Iron.   Cementite (White) with Lamellar Pearlite ($\times 5oo$).

GRAPHITE (Fig. 179): concerning its formation see p. 188.

TEMPER CARBON is non-graphitic C which separates out from white iron by keeping it for a long time at a temperature near 1000°, during which time the finely-divided cementite changes into a mixture of ferrite, pearlite and temper carbon (Fig. 180).

This form of carbon is more readily oxidizable than graphite or carbide carbon (cf. Malleableizing, p. 198).

2. SLAG.   $SiO_2$ and $Al_2O_3$ are almost invariably present in the

gangue of iron ores and in the ash of the fuel. Again, the oxides of the alkaline earth metals, particularly of lime, are always present in the slag, since CaO is used as a slagging flux (cf. p. 181). The chief constituent of blast furnace slag, therefore, is a calcium-aluminium silicate, and this silicate is approximately neutral or weakly basic, according to the impurities in the raw materials and according to the desired nature of the iron to be produced, and it is more or less rich in other metallic oxides (FeO, MnO, etc.).

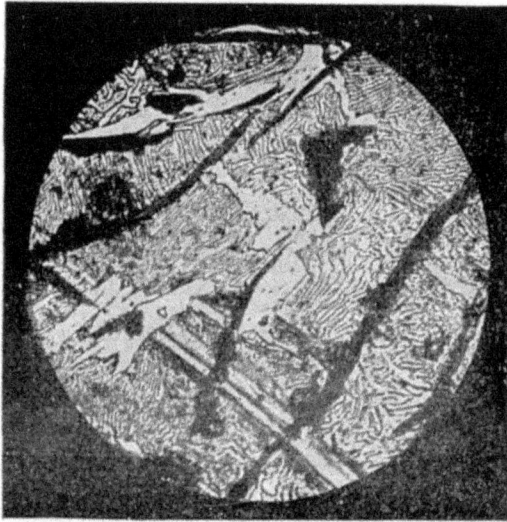

FIG. 179.—Gray Cast Iron   Graphite (Black);   Cementite (White) with Lamellar Pearlite ($\times 500$).

The basic slags crumble readily after exposure to the atmosphere, but when it is desired to utilize the less basic slags they are usually granulated by allowing them to flow into water as they are tapped from the furnace. The more basic slags, particularly of foundry iron, when they contain suitable amounts of $SiO_2$, $Al_2O_3$ and CaO, are used in the preparation of Portland cement.

3. FURNACE GAS. In discussing the reactions of the blast furnace, it was mentioned that a gas escapes from the

top of the furnace containing up to 28% of CO (p. 187). One cubic foot of such a gas possesses a heating value of 89 B.T.U. For one ton of pig iron, there are about 160,000 cu.ft. of this gas. A part, half at the most, serves to heat the blast that is to be used in the furnace, and there remains from 80,000 to 90,000 cubic feet of the gas for other purposes. Until recently this gas was used for heating boilers which served as a source of power for

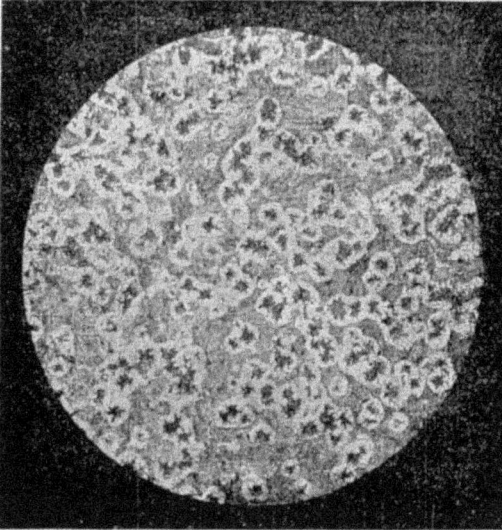

FIG. 180.—Cast Iron showing Temper Carbon. (Black); in Ferrite (White), Surrounded by Lamellar Pearlite.

moving the raw materials and the products of the furnaces; to-day the gas is being utilized chiefly in gas engines. The blast-furnace industry, however, requires but a quarter of the power that can be produced thus, and if this additional power is not needed for other work in the plant (e.g., for rolling mills) it can serve for outside purposes (e.g., for electric plants). The amount of power thus available can be easily computed. After making the deductions as just outlined for the gas that is utilized in connection with the blast-furnace work, there remains 75% of 80,000 cubic feet of gas, or from a 300-ton furnace, $60,000 \times 300 = 18,000,000$ cubic feet of gas in 24 hours; 125 cubic feet of the gas yield approximately 1 horse power hour. Thus 18,000,000 cubic feet from a 300-ton furnace will yield hourly about $\dfrac{18,000,000}{125 \times 24} = 6000$ horse power hours.

# (C) Iron Alloys and Iron Compounds

Besides pig iron, certain alloys of iron and iron compounds to be used in the preparation of different kinds of forgeable iron and steel are obtained by smelting in blast furnaces.

**Spiegeleisen and Ferromanganese,** see Manganese.

**Ferro-Chrome,** see chromium.

**Ferro-Molybdenum** is obtained in the process of Borchers and Lehmer by smelting molybdenite ($MoS_2$) in an electric furnace with lime, carbon, and Fe or a sulphide or oxide of iron.

**Ferro-Vanadium** is prepared by a reducing fusion of ferrous vanadate with carbon and enough Fe in the flux so that the resulting alloy will contain not more than 25% Vd. The ferrous vanadate is obtained from sodium vanadate, which is prepared by an oxidizing roast of vanadium ores with $Na_2CO_3$ or NaCl, or by fusing lead vanadate with soda; the sodium vanadate is extracted with water and the clarified solution is treated with $FeSO_4$ solution.

**Ferro-Silicon** can be produced like pig iron in the blast furnace by smelting iron ores with a gangue or flux rich in $SiO_2$. In such cases the aim is to produce a slag rich in $Al_2O_3$. By the blast furnace method the Si content may be made as high as about 16%. Richer alloys, with 25%, 50% or more of Si, are usually obtained in electric furnaces by the reduction of mixtures of iron ores and sand. Ferro-silicon has also been obtained as a by-product in the manufacture of $CaC_2$. In this case a coal with an ash high in $SiO_2$ was used. Iron ore was then added to the charge and a ferro-silicon with about 20% Si obtained.

# (D) Forgeable Iron

The equilibrium diagram, Fig. 176, p. 190, contains a point $a$, which corresponds to 2% C, and represents a solid solution of the composition $2Fe_3C + 15Fe$. This is the saturation point of $\gamma$-iron for $Fe_3C$ and represents, at the same time, the upper carbon limit for the forgeability of iron. Although in practice the upper limit for a forgeable iron is taken as 1.6% C, still in the light of the relations between Fe and C which have been sufficiently cleared up scientifically, we may regard as forgeable iron all alloys lying between pure Fe and the saturated solution of $2Fe_3C$ in 15Fe.

At $S$ the C content amounts to 0.9%, corresponding to a composition of $Fe_3C + 20Fe$ (pearlite). Although with a C content considerably lower than 0.9%, there is an increase in hardness when the metal is heated and then rapidly cooled, this increase in hardness by such treatment begins to be very marked at 0.9% C. At this point a solution is reached from which, on slow cooling, pure Fe (ferrite) can no longer crystallize out as an independent constituent of the alloy, but is contained only in the eutectoid called pearlite, $Fe_3C + 20Fe$; with higher C content the eutectoid is formed in the presence of $Fe_3C$.

The characteristic hardening of steel has been explained by assuming a transformation of pearlite into martensite by heating and maintaining this solid solution by quenching; on the other hand, the transformation of steel into the malleable condition has been attributed to a transformation of martensite into pearlite, either by reason of slow cooling or by keeping the quenched steel at moderate temperatures (below 700°) for a long time. The more the C content falls below 0.9%, the more ferrite exists in the presence of pearlite. It is, however, easy to understand why a steel with less than 0.9% C can be hardened. When 0.9% C is present, the sample can contain 100% of pearlite, and therefore a steel with 0.3% C contains 33.3% of pearlite which can be transformed into martensite. On the other hand, it is plain that the hardening power and malleability will not disappear when the C content is above 0.9%, for beside the cementite, which now occurs in ever-increasing amount, it is still possible for pearlite to be formed and it is not until 2% C is reached that the amount of pearlite is so small in comparison with the amount of cementite present that the malleability practically ceases.

The property of hardening can be caused with very low C content by the presence of other substances such as, for example, Cr and W.

The transformation of pig iron into forgeable iron must evidently consist in diminishing the amount of C present and removing as completely as possible the other undesirable constituents of the crude iron. In all cases this is accomplished by oxidation processes. In one case (malleableizing) only C is removed, but in all other cases, other constituents are removed simultaneously. At the same time it is impossible to prevent a part of the Fe itself from being oxidized and passing partly into the slag and partly into the purified iron as FeO, so that a further reduction is necessary to remove the oxide.

**Malleableizing.** In 1903 to 1905 F. Wüst and some of his students proved experimentally that by exposing white iron for a long time to a temperature of about 1000° (annealing) there is a decomposition of the dissolved $Fe_3C$ into $3Fe + C$ in such a way that the C is not deposited as graphite, as in the formation of gray iron, but entirely in an amorphous condition such that it is readily susceptible to oxidation. As oxidizing agents, oxide iron ores, usually $Fe_2O_3$, but sometimes $FeCO_3$, are employed. The iron objects to be malleableized, already in the form of finished castings, are packed in this ore, gradually brought to a temperature lying between 900° and 1000° and kept there about a week. Until recently it has been held in metallurgical literature that the oxidation of the C begins at the surface where the oxidized ore is directly in contact with the iron, and that then the solid C migrates from the interior to the carbon-free outer layers, where it is oxidized in turn. The author, in his Inorganic Chemistry, opposed this view in 1893; in 1908 Wüst submitted indisputable experimental proof that the decarburization is brought about by the aid of $CO_2$, which is obtained partly by the contact of particles of iron containing C with the $Fe_2O_3$, and partly from the combustion gases of the furnace. Wüst showed, moreover, that these gases diffused into the interior of the castings, where the most favorable conditions exist (cf. diagram 174) for the progress of the reaction $CO_2 + C = 2CO$, while the CO thus formed is again converted into $CO_2$ by the oxidized ore, and can once more take part in the reaction.

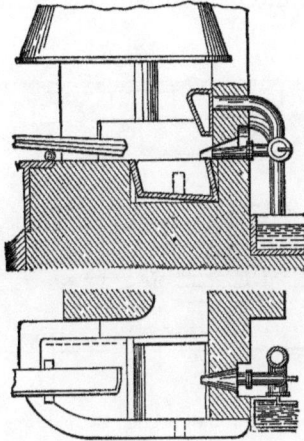

FIG. 181 —Charcoal Hearth.

**Production of Wrought Iron.** These methods of refining iron are the oldest in the development of iron metallurgy.

Among them are:

**The Charcoal Hearth Process.** This is still used in some places (Sweden, Styria, Russia) where a sufficient supply of charcoal can be obtained cheaply. The " hearths " or " forges " contain a hollow pit, usually lined with iron plates, and are provided with a tuyere in the high back wall. The cast iron is melted upon charcoal but with a blast of air sufficient for the necessary oxidation, and directed so that the liquid iron must drop through the blast zone. A glance at the diagram, Fig. 176, shows that the melting point of iron rises as its purity increases. The iron, that was at first liquid, collects in the bottom hearth as a pure viscous metal (blooms) which is freed from any included slag by mechanical working (hammering, rolling), and welded to a compact mass.

**Puddling.** This is essentially the same kind of work, but differs by melting the cast-iron in a reverberatory furnace, in which the oxidation is assisted by the presence of iron oxides, added in the form of slags rich in $Fe_3O_4$, and waste from the mechanical working of forgeable iron (e.g.,

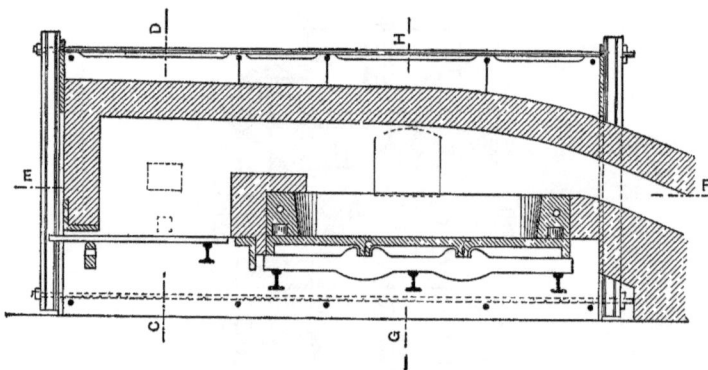

FIG. 182.—Longitudinal Section of Puddling Furnace.

hammer scale). The contact of these oxides, and of the O from the flame gases, is aided by stirring up the slag with rabbles. As the pure iron solidifies it is broken up from time to time and turned over, etc., so as to expose the whole charge to the refining action. As in the last-mentioned process, the metal is finally freed from slag by hammering, and is thereby welded together into a conglomerate of Fe crystals. The reverberatory furnaces consist of beds or hearths supported by iron plates, with a free space underneath so that they are exposed to the cooling action of the atmosphere. The side walls are hollow iron plates, which can be cooled with water, and are lined with a basic slag which has a

high melting point, and is rich in FeO. The older, single-
hearth reverberatory furnaces have a large grate area and
the waste gases which pass off at a temperature of 1200 to
1300° C. are used, as a rule, for making steam. The
newer furnaces have regenerative gas firing and double
hearths. According to the requirements placed upon the
iron, 10 to 20 charges of 500 to 700 lbs. each are worked
up in a day in the older furnaces; in the newer ones, 24
charges of 1000 lbs. each. Loss of metal, 6 to 15%.
For heating the furnaces and for power sufficient to hammer

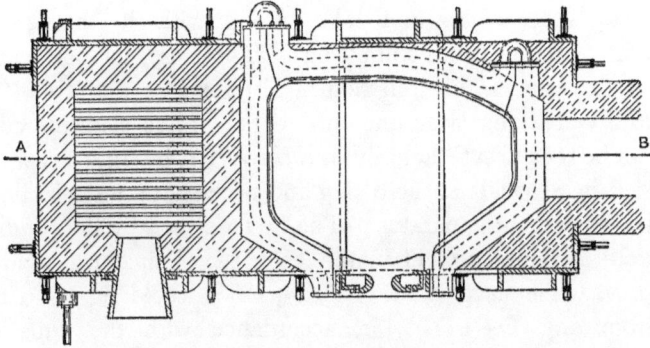

Fig 183 —Plan of Puddling Furnace with Flat Grate, burning Anthracite Coal

or roll the blooms, the coal consumption equals the weight
of the finished product.

The forgeable iron produced by either of the last two processes
is called *wrought iron*. Since the temperature of the
furnace hardly reaches 1300°, it is evident that the crystals
of iron which are formed during the refining cannot fuse
together but become welded.

**Oxidizing Fusion of Cast Iron Above the Melting-point of
Pure Iron. Production of Steel.** The oxidation is accomplished by atmospheric oxygen, as well as by the oxygen of iron
oxides, by blowing in converters; or by smelting in regenerative furnaces, or in electric furnaces.

The **blowing** consists in conducting a blast of air through the molten cast iron until the impurities have been burned away; the heat of combustion serves to maintain a temperature so high that the purified iron remains liquid. The substances present have the following heats of combustion per pound: $Si = 14,090$ B.T.U.; $P = 10,740$ B.T.U.; $C = 4297$ B.T.U.; $Mn = 3103$ B.T.U., and $Fe = 2435$ B.T.U. The rise in temperature caused by these elements is in the following order: By the combustion of $1\%$ of the entire weight of the charge, Si increases the temperature, according to recent calculations of Wüst, $287°$ C.; the same amount of P, $185°$ C.; of Mn, $61°$ C.; of Fe, $44°$ C.; and of C., $8.8°$ C. (CO escapes as a gas and takes along with it the greater part of the heat of combustion). Therefore, for purification by this process, an iron rich in Si or in P is desirable. An iron containing large amounts of both these elements could not be refined satisfactorily in a converter, because a cast iron rich in Si yields an acid slag and necessitates an acid lining in the melting apparatus. The $SiO_2$ formed by the combustion, and that present in the converter lining, will tend to prevent the slagging off of $P_2O_5$, for it would set $P_2O_5$ free from any $Fe_3(PO_4)_2$, in accordance with the equation:

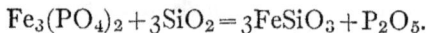

$$Fe_3(PO_4)_2 + 3SiO_2 = 3FeSiO_3 + P_2O_5.$$

Free $P_2O_5$, however, is readily reduced by the large excess of hot iron present and the P will thus pass into the iron again in the form of phosphide. The process, therefore, is conducted according to one of two principles, the older one being named, after its discoverer, the

**Bessemer Process.** In this case pig irons high in Si (up to $2\%$ Si) are " blown " in tilting, pear-shaped vessels called Bessemer converters. These are made of wrought-iron plates bolted firmly together and are lined with an infusible silicious rock termed ganister. They are fitted with tuyere boxes at the bottom. In the

**Basic Bessemer** or **Thomas-Gilchrist Process,** a cast iron rich in P is used (up to $2.5\%$ P) with as low Si as possible. The

converters, in this case, are given a basic lining consisting
of calcined dolomite with the addition of lime. The slags
formed consist of a basic calcium phosphate, with $P_2O_5$
up to 25%; this forms a valuable by-product, the so-called
*Thomas slag*, which is used as a fertilizer.

The converters are built in different sizes, up to 10 ft. in
height and 10 ft. in width (measuring the shell without
lining). They will hold from 1.5 to 25 tons of pig iron,
which is blown in from 15 to 20 minutes. The lost metal
amounts to 10 to 15%. As adjunct to the Bessemer plant,
there remain to be mentioned the

PIG-IRON MIXER, which is a revolving drum on a horizontal
axis; it is made of iron plates with a silicious lining, and
has a capacity up to 600 tons. By this mixing, a fairly
uniform material is guaranteed for the converter and a
certain preliminary purification takes place, inasmuch as
Mn combines with S and the MnS separates out as slag.

**Open Hearth Process.** This process was rendered possible by
the discovery of regenerative firing by Friedrich and Wilhelm
Siemens and first utilized by Emil and Pierce Martin in a
Siemens furnace for the production of steel (1865).

In the **Siemens-Martin Process** the oxidation of the impurities
in the iron is effected partly by the oxygen of the air
and partly by the oxides added, which is either rust on scrap
iron, or in the form of hammer-scale, or pure ore (magnetite,
hematite). According to whether the raw material is rich
in Si or in P, the work is either carried on " acid " or " basic,"
as in the Bessemer process, and the lining of the hearth
is governed in the same way.

While the Siemens-Martin process was formerly used
almost exclusively for working up pig iron or wrought
iron into steel, it has since been used with much success for
smelting pig iron with iron ores; or, in other words, a method
of working has developed in which considerable amounts
of steel are obtained directly from ores. Such processes
are the recent ones of **Bertrand-Thiel** and of **Talbot.** In

the former, the work is carried out with two Siemens furnaces with basic hearths in such a way that in one furnace a preliminary refining takes place with the production of a slag

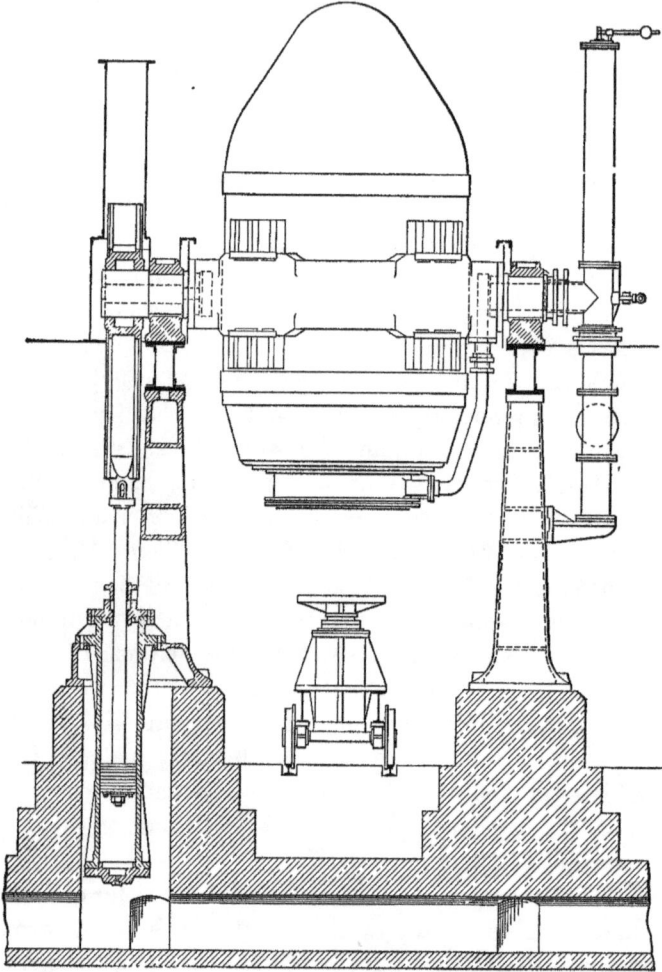

Fig. 184a.—Converter.

rich in $P_2O_5$, and in the second the finishing work is carried out with a fresh addition of ore. The Talbot process works with one large furnace (up to 250 tons charge) in which the liquid pig iron is run upon the previously-introduced ore and lime.   After the conclusion of the violent reaction that takes place, the first slag is removed in order to smelt the metal further with another addition of ore.

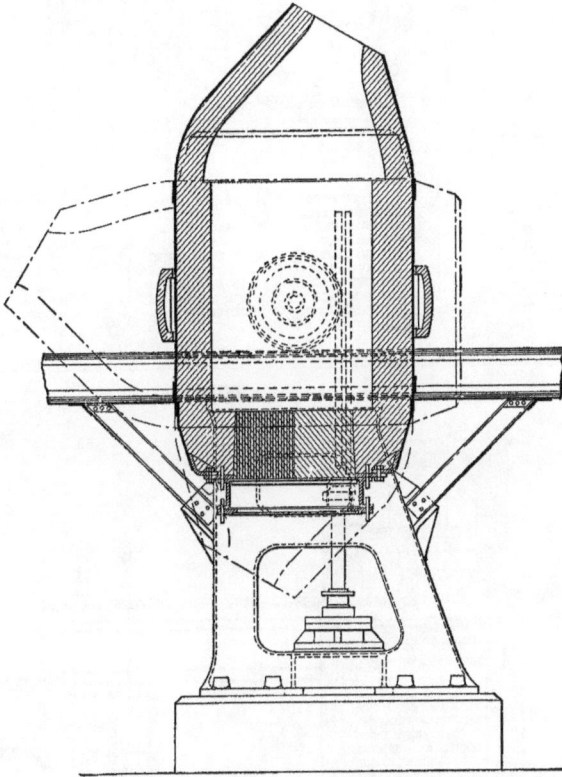

FIG. 184b.—Converter.

APPARATUS: Reverberatory furnaces with open hearths and with regenerative firing. The hearth rests upon iron plates and consists either of lump quartz or of calcined dolomite. The other walls of the smelting space are also of acid or basic material, at least on the inside. The side walls contain apertures for working, charging, and tapping; the end walls have openings for admitting the

gas and air which are alternately introduced, from one
side and then the other, after passing through the heating

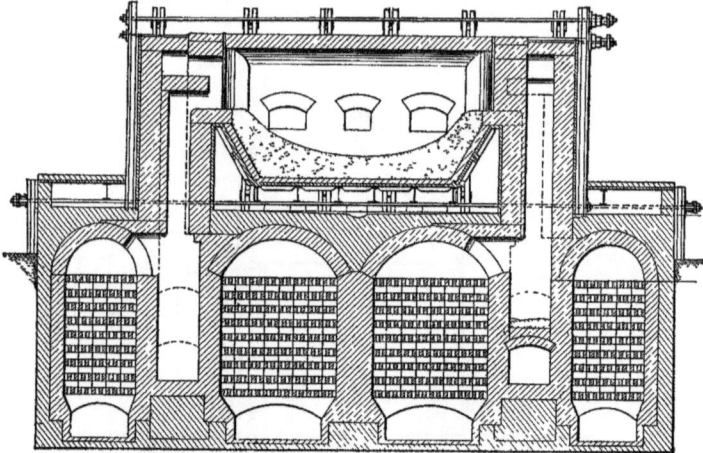

FIG. 185.—Siemens Open Hearth Furnace.

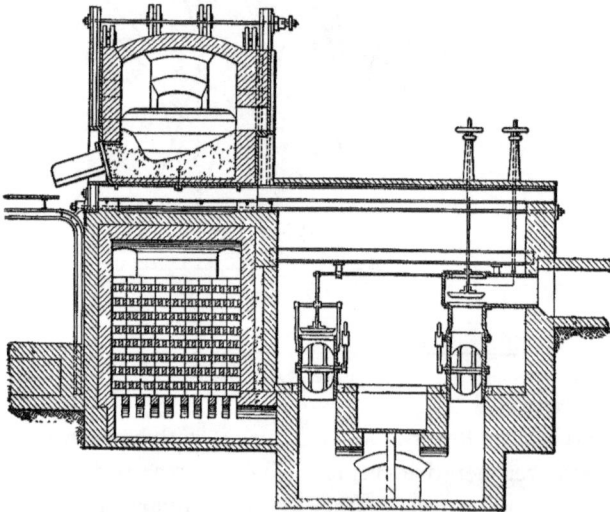

FIG. 186.—Siemens Open Hearth Furnace.

chambers lying beneath the hearth. Each furnace, therefore,
has two pairs of heating chambers which consist of shaft-
like structures latticed with firebricks to aid in the trans-

ference of heat. One chamber of each pair serves for heating the combustible gases, and the other for heating the air required in the combustion of these gases. During the operation of the furnace, therefore, the current of gas is sent into one chamber from the bottom, and the air is sent into the other chamber. Both of these currents pass through the narrow openings in the brickwork and are heated by contact with the hot bricks. The two streams are kept separate until after they leave the top of the chambers, when they are introduced into the smelting space and combustion at once takes place. Since a high temperature is required in the smelting zone, the heat transference cannot be carried very far here. Hence, the hot waste gases, as they leave this zone, are led through channels in the opposite wall, into the second pair of heating chambers in which, as they pass downward and out at the bottom, they heat the bricks and walls. After an interval of 30 to 50 minutes the direction of the gas currents is reversed and the air and combustible gases are now conducted through the chambers that have just been heated by the escaping gases.

The furnaces are sometimes made so that they will tilt and are then suited for very large charges. The ordinary furnaces with fixed hearths have a capacity of from 12 to 40 tons, but the tilting furnaces will hold up to 300 tons. From three to six charges are worked through in 24 hours. Metal losses, 6 to 8%.

**Bessemer and Open Hearth Treatment in Succession** (Duplex Process) has been practiced in some places (e.g., in Wittkowitz) for working up a pig iron containing too much P to be satisfactorily purified by the ordinary Bessemer process, and too much Si for the basic Bessemer process.

**Electric Smelting,** like the Siemens-Martin process, can be used for working up either scrap or pig iron. In 1878, soon after the discovery of the dynamo, Wilhelm Siemens proposed to smelt iron by electricity, and since that time there has been

no end of experiments toward solving the problem, but up
to 1900 these were mostly concerned with the question of
electrodes. In heating by the electric arc, carbon electrodes
are indispensable. When, however, the electric arc springs
from a carbon electrode to the mass to be heated, it is unavoidable, if this mass consists of a metal like iron, that
some carbon will be taken up by it. The taking up of
carbon, in spite of the use of two carbon electrodes, was first
prevented satisfactorily by means of the

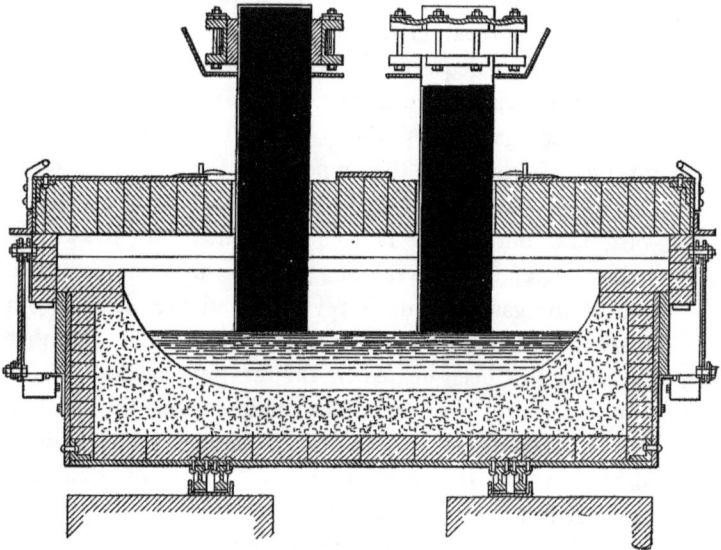

FIG 187.—Héroult Furnace.

HÉROULT FURNACE, consisting of an electric arc and resistance furnace in which both electrodes are introduced from
above into the smelting hearth. An oxidizing slag (magnetite and a basic flux) is kept upon the metal. The
electrodes are placed far enough apart, and the layer of
slag on the other hand is kept so thin that the current
passes from one electrode, with the formation of an arc,
first to the slag, then to the metal, from the other end of
the metal again to the slag, and thence to the other elec-

trode with the formation of a second arc. Evidently, then, as sources of heat there are: two electric arcs playing directly upon the surface of the slag, two layers of slag resting directly upon the metal, and the metal itself as heating resistance. The slag, containing the substances capable of entering into chemical reaction, is heated very hot and so is the metal itself, which is well protected from loss of heat. The C from the electrodes is oxidized in the slag layer. A glance at the whole arrangement shows, and this has been confirmed in practice, that the oxidizing agent in the refining slag is brought by the intense heat to its maximum reaction velocity in respect to the impurities present in the bath of iron. As a matter of fact, it is possible to accomplish in this furnace, and in that of the following ones, a much more efficient purification of the iron than in any of the previously described furnaces. The potential required for the Héroult furnace is 110 to 120 volts.

GIROD FURNACE. This arc-resistance furnace was designed by Paul Girod-Ugine and, as compared with the Héroult furnace, excels in simplicity of construction and operation. One pole of the electric arc, consisting of one or more blocks of carbon, is introduced from above into the middle of the smelting hearth. The slag floating upon the metal forms the other pole. It obtains its contact through the metal bath, which is placed in circuit with iron rods, or rings, that are introduced from below into the furnace near the periphery of the metal bath. These connecting pieces, which are cooled somewhat outside the furnace, are so dimensioned that they can just conduct the current without too great resistance; in other words, they are given as much load as possible so that there is a good distribution of the current all over the bath. The ends, drawn out somewhat where they are in contact with the bath, are kept just hot enough by the current to prevent any undesired conducting away of heat from the bath. As sources of

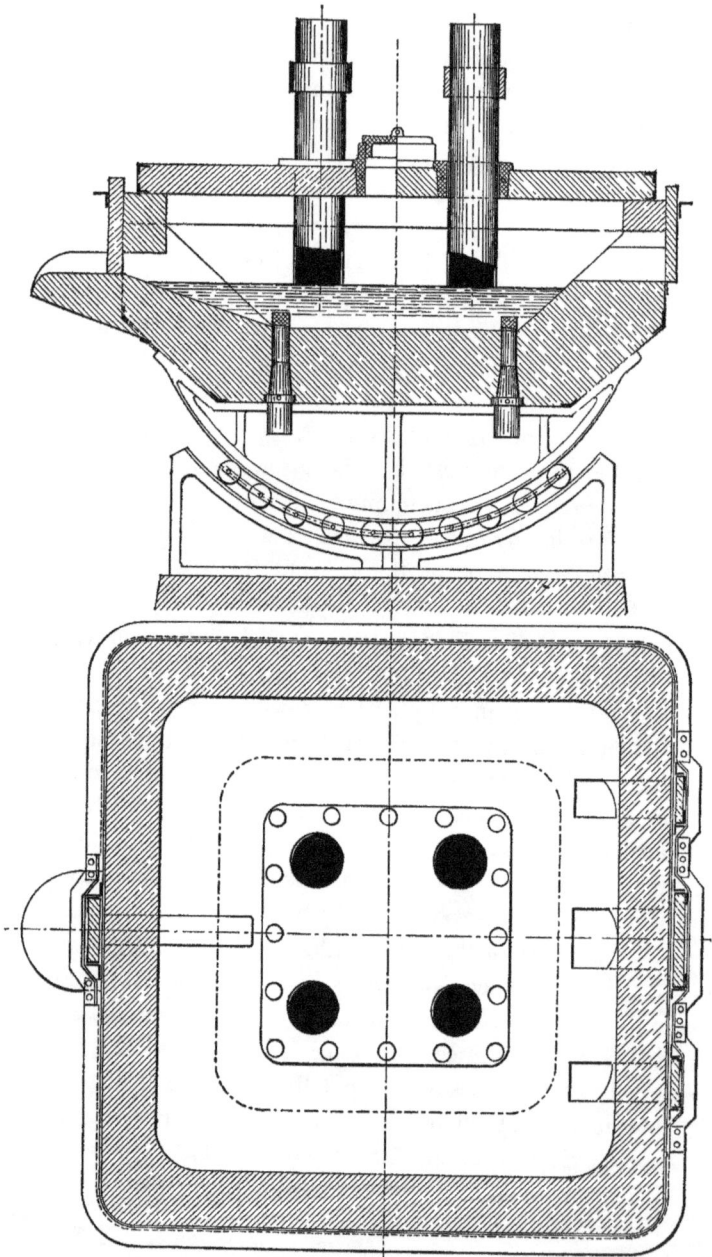

FIG. 188.—Girod Furnace.

heat in the Girod furnace, therefore, there are: one or more electric arcs over the middle of the bath, a slag layer upon the metal, and the metal itself. The current passes with the formation of an arc from one or several upper electrodes through the refining slag to the middle of the metal, whence it radiates uniformly in all directions, heats the bath, and keeps it in motion, and is carried away in the outer periphery. For the Girod furnace, a potential of 55 to 65 volts is necessary.

INDUCTION FURNACES, a discovery of de Ferrantis in the year 1887, were first adapted by Kjellin, and almost at the same time by Colby (1900), for use as furnaces for iron smelting. A closed electro-magnet, built in a right angle, forms a transformer with the fused, or ready-to-be-fused, iron which rests in a ring-shaped gutter that is placed perpendicularly to the iron core of the magnet. The most effective construction has proved to be that in which the winding of the electro-magnet lies concentric to the iron-bath, whether this be within or without the field of the primary circuit. The secondary current produced in the iron of the gutter by an alternating current in the primary circuit is at once transformed into heat and the iron is quickly melted. Since the furnace works without electrodes, there is no danger of contamination from electrode material, but notwithstanding this, the majority of these induction furnaces have not proved satisfactory. The oxidizing slag (iron oxides, etc.) owing to its low conductivity, takes but little part in the formation of the secondary current and in the resulting heat transference, but obtains its heat from the iron. The slag, therefore, is considerably colder than that of the Héroult and Girod furnaces. In those furnaces, the slag acts as one pole of the arc and, because of its poor conductivity, causes the production of much heat in the passage of electricity through it, and this high temperature is particularly favorable for increasing the velocity of the reactions that

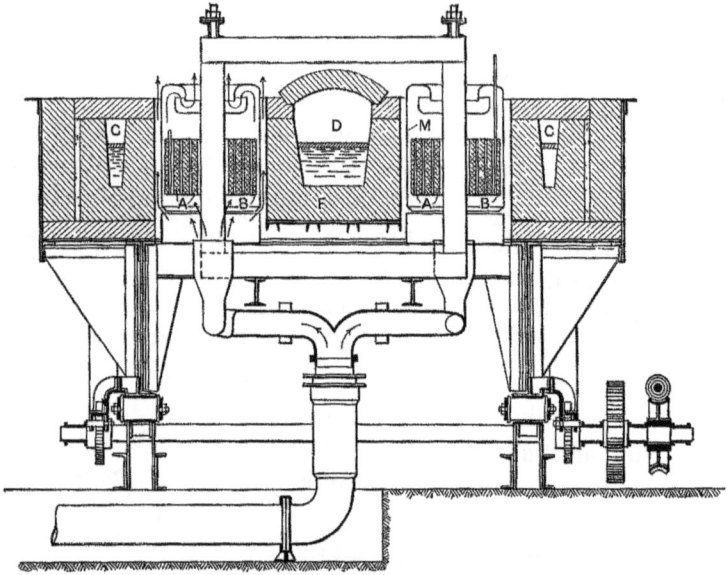

FIG. 189.—Ròchling-Rodenhauser Furnace.

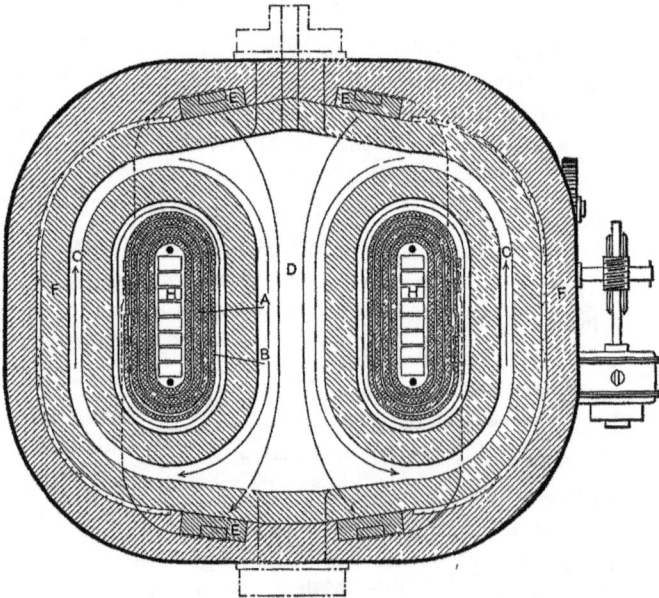

FIG. 190.—Rochling-Rodenhauser Furnace.

tend to take place between the oxygen of the slag and the impurities of the iron. This difficulty is overcome in the furnace of RÖCHLING and RODENHAUSER by giving direct resistance heating to a part of the bath. Both vertical arms of the magnet's iron core $H$ (Fig. 191), are provided with windings, $A$, of the primary circuit, and with melting gutters $C$; these last unite in a broad middle hearth, $D$. At the ends of this middle hearth are two protective walls

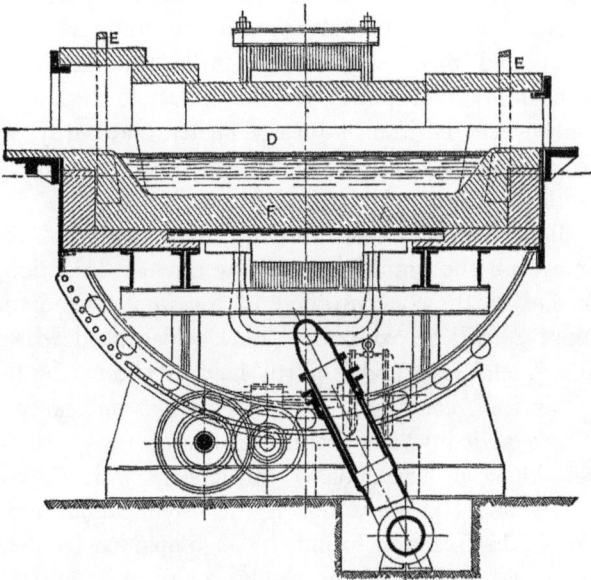

FIG. 191.—Röchling-Rodenhauser Furnace.

of magnesia or dolomite, both of which materials conduct well at high temperatures. Back of these walls and connected with a secondary current $B$, are pole plates $E$ from which currents are said to pass through the charge contained in the broad hearth.

CAPACITY OF ELECTRIC FURNACES. Whereas the Héroult and Girod furnaces work advantageously even with cold material, the induction furnaces are practically restricted to work with molten and partly purified iron. Starting with cold metal as raw material, in a small Héroult or Girod furnace (2 to 2.5 tons charge) about 900 kilowatt hours = 1230 horse-power hours are to be reckoned per ton of finished steel; in large furnaces (8 to 10 tons), 700 kilowatt hours = 951 horse-power hours. In the first case, therefore, about 0.14, and in the second 0.108 horse-power year per ton of steel. If the furnace can be charged with liquid metal, then the expenditure of energy is diminished by at least one-third. The Girod method of introducing the electric current gives greater velocity and uniformity when working with cold metal than does the Héroult method. The consumption of power by the induction furnace is less only when the iron is already melted and partly purified; when cold metal is used, more power is required than with the Héroult and Girod furnaces.

In all processes for the production of steel, it is possible to remove all the impurities only by an over-oxidation, so to speak, i.e., as the concentration of the impurities diminishes, the more Fe will be oxidized to FeO, which partly passes into the slag (metal loss) and is partly held in solution by the iron. If to this iron, containing FeO, the various substances C, Si, etc., were added which are necessary for the production of special kinds of steel, then, particularly with C, chemical reaction would take place during the solidification, and in the case of C this reaction would be accompanied by the evolution of a gas (CO) and the result would be a porous metal. FeO, remaining dissolved in the iron, makes the metal *red-short;* (i.e., brittle when hot). After the fusion is completed, therefore, there is in all cases a

**Reduction** by the addition of Mn in the form of a Mn-Fe alloy, of C for the recarburization, of Si in the form of silicides, and finally of Al for the removal of the last traces of FeO and for accelerating the reactions between FeO and the other substances charged. The addition of the Al takes place usually in the ladles, or even after pouring into the molds.

**The Finishing of Forgeable Irons.** In the refining methods which have been described, it is practically impossible to stop the process when the metal contains the desired amount of C or Si corresponding to a certain quality of iron. The impurities, are, therefore, removed completely and then substances are added which will impart to the iron the desired properties. The simplest and quickest way to do this, is by

**Alloying in the Bath,** which is accomplished in the production of steel by adding the necessary substances themselves to the bath of pure Fe, or in the form of an alloy or chemical compound, either in the smelting furnace or in the ladle. For the recarburization of iron up to a certain definite point, it is sufficient to add pure coke or graphite to the ladle as the iron is poured from the converter (Darby-Phoenix Process). It is more usual, however, to add spiegeleisen (see Mn) with high C content. Similarly, for introducing Si, a ferro-silicon is added. Also for introducing other metals, such as Mn, Cr, W, etc., it is customary to add these in the form of an iron alloy on account of its being easier to prepare such an alloy of sufficient purity than the pure metal itself; W and Cr as well as Ni and other metals, are also added in the pure state.

For producing varieties of steel of very uniform composition, particularly when it is not necessary to keep the C content low, the pure, analyzed metal is melted with the required additions (Ni, W, Cr, etc.) in graphite crucibles. In this way *crucible steel* is prepared, and recently the melting has been done in an electric furnace.

**By Welding,** accomplished by pressing together (hammering or rolling) rods of different kinds of iron at a welding heat, it is possible to prepare different qualities of iron and steel which apparently are uniform but when examined under the microscope show that different materials have been united mechanically. Particularly for the so-called *double shear steel* (for cutlery, etc.) it is important to have hard particles embedded in a softer and sufficiently elastic matrix; in this way the sharpness of instruments is maintained.

**Cementation.** This process, like that of malleableizing, retains the form of the objects subjected to it, but the nature of the process is exactly the reverse. The objects are embedded in charcoal powder and heated for a long time to a temperature of about 900° C.; in this way the C content can be raised to about 1.2%. The working conditions are favorable for the reversal of the reaction that takes place during malleableizing; $2CO = C + CO_2$. The $CO_2$, as fast as it is formed, is reduced by the bed of C to CO again and can serve anew for the introduction of C into the iron. It is also possible to carburize iron objects superficially by dipping them in baths or exposing them to the action of certain gases. These compounds, or gases, on being heated, give rise to C or carbides. The process, called *case hardening*, is accomplished by heating the metal and plunging it into some substance like yellow prussiate of potash (potassium ferrocyanide) $K_4Fe(CN)_6 = 4KCN + FeC_2 + N_2$, or by heating in a current of illuminating gas.

## Properties of Iron:

SPECIFIC GRAVITY: 7.86.

COLOR: grayish-white with high luster.

MECHANICAL PROPERTIES: tough, very ductile.

STRUCTURE: see ferrite, page 192, Fig. 177.

MELTING-POINT: 1512° C. (2754° F.)

BOILING-POINT: 2600° C. (?) (4712° F.)

ELECTRICAL CONDUCTIVITY: about 0.14 that of Ag.

MAGNETIC PROPERTIES: Fe is the most paramagnetic of metals.

ALLOYS readily with most of the earth metals, slightly with Pb and Cu. In the presence of Si, Fe will take up more Cu; or, in other words, it dissolves copper silicide more readily than it does pure Cu. Fe also alloys easily with its own compounds with metalloids (C, Si, P, S, O) (Cf. pig-iron, page 188).

CHEMICAL BEHAVIOR: in dry air it is fairly stable toward
O at low temperatures. At 300° C. it is more readily oxidizable. At a moderate red heat it combines with S and P,
and at higher temperatures with C and Si. The halogens
act energetically upon Fe even at ordinary temperatures.
On account of its high electrolytic solution tension ($+0.344$
toward H) it dissolves easily in dilute mineral acids
with evolution of H, and in $HNO_3$ with reduction of the
acid and evolution of nitric oxide. In concentrated
$H_2SO_4$ or $HNO_3$, it is so difficultly soluble that iron vessels can be used for transporting these acids. In the technically-important iron compounds, the Fe is either bivalent
(ferrous compounds) or trivalent (ferric compounds).
Only in the carbides and silicides, which are formed and
are stable at high temperatures, does Fe show a different
valence. The most important carbide has the composition $Fe_3C$ (cementite), and silicides are known corresponding to the formulas $Fe_2Si$, $FeSi$, $Fe_3Si_2$, and $FeSi_2$.

The properties of the technically-important varieties
of iron have already been discussed sufficiently in the
sections defining the terms *cast iron* (white and gray),
*wrought iron*, and *steel*.

# CHROMIUM

## Sources

**Natural Sources:**

CHROME-IRON ORE, or chromite, $FeCr_2O_4$ ($FeO.Cr_2O_3$), usually accompanied by serpentine. This is the most important chromium ore. Of the other minerals, the best-known is

CROCOITE, lead chromate, $PbCrO_4$, now scarcely used at all for the production of metallic chromium.

## Ferro-Chrome

**A Reducing Smelt** of the chrome-iron ore gives directly a commercial Fe-Cr alloy. As flux, only charcoal or coke powder is required. At the beginning of the reaction, which takes place with evolution of CO, there is danger of the mixed ore and carbon falling apart on account of the difference in specific gravities; to prevent this, a little colophonium or pitch is added to the charge.

On being heated, the pitch glues together the ore and carbon, forming a coke at a high temperature, and then finally the whole mass unites to one large lump. The charge is regulated, of course, by the nature of the ore. To one ton of ore, from 250 to 330 lbs. of charcoal powder and 125 to 155 lbs. of powdered colophonium or pitch is added.

APPARATUS: Crucibles in reverberatory furnaces, using forced draft or regenerative firing, or in electric furnaces.

THE CRUCIBLES are made of graphite or of clay. When many of the former material are required, the cost of operating is materially increased. Clay crucibles can be used but once,

but they are cheap. In the simple furnaces the consumption of fuel is very great and it is hard to reach the
necessary temperature. In this respect the regenerative
furnaces are more advantageous. Since in using clay crucibles
the heating chambers must be cooled enough so that
there is no danger of breaking the crucibles in recharg-
ing the furnace, the author usually combines two furnaces

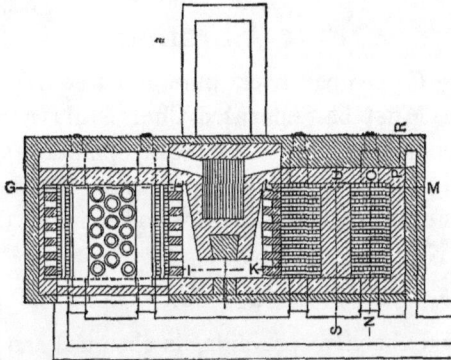

Fig. 192.—Section A, B, C, D, E, F.

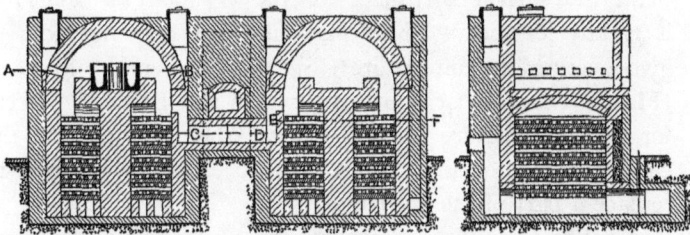

Fig. 193.—Section G, H, K, L, M.      Fig. 194.   Section N, O, P, R.
Figs. 192-4.—Crucible Furnace with Regenerative Chambers.

in such a way that it is possible to work alternately, the
furnace with the fresh charge receiving its heat from the
slowly-cooling furnace with the finished charge. The furnace
block contains the gas producer and the two regenerative,
reverberatory furnaces. (Figs. 192 to 194.) The producer
is a simple shaft furnace with horizontal grate for coke.
It sends its gas into one of the main flues of the adjacent
reverberatory furnace. From the main flue, the gas can be

led into one or the other branching flues of each furnace by uniting two faucet tubes with a ∩-shaped tube of sheet metal. The regenerators are there only for preheating the air. The work is carried out as usual, changing the direction of the gas every half hour or hour. The change in direction of the producer gas is effected by the above-mentioned faucet tubes, of the air and waste-gases by valves.

# Pure Chromium

To prepare Cr comparatively pure and free from Fe,

**I. The Iron Must be Separated Chemically** from chromium. This is accomplished by the following operations:

**1. Oxidizing Roast with Alkaline Fluxes.** Purpose: changing FeO to $Fe_2O_3$, $Cr_2O_3$ into $CrO_3$ and union of the latter with CaO or $Na_2O$.

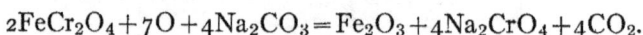

$$2FeCr_2O_4 + 7O + 4Na_2CO_3 = Fe_2O_3 + 4Na_2CrO_4 + 4CO_2.$$

Contrary to the views prevailing in chemical and metallurgical literature, the *mass should not melt*, for when melting takes place the surface of attack for the O upon the $FeCr_2O_4$ is greatly lessened. The firing, therefore, does not need to give a very high temperature. Simple reverberatory furnaces (Figs. 195 to 197) such as are used in the LeBlanc process for the manufacture of soda, are suitable. To prevent the charge fusing together, $CaCO_3$ has been added or used to replace a part of the $Na_2CO_3$.

**2. Lixiviation.** Purpose: separation of soluble chromate, $Na_2CrO_4$, from insoluble $Fe_2O_3$ and gangue. As solvent, hot water is used and if $CaCrO_4$ is present, $Na_2CO_3$ or $Na_2SO_4$ is added. In this case, the work is performed in closed iron drums at 120° to 130° C. The filtered $Na_2CrO_4$ solution is evaporated to a concentration of 1.5 sp.gr.

**3. Transformation of $Na_2CrO_4$ into $Na_2Cr_2O_7$.** If $H_2SO_4$ is added to the concentrated, aqueous solution of the chromate the following reaction takes place:

$$2Na_2CrO_4 + H_2SO_4 = Na_2Cr_2O_7 + H_2O + Na_2SO_4.$$

The $Na_2SO_4$ is deposited, to a large extent, as a crystalline powder. The liquor is drawn off from the sediment, and the solution concentrated, whereby practically all of the remaining $Na_2SO_4$ separates out. If care was taken to

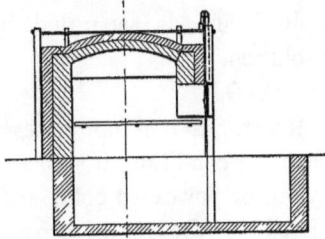

FIG. 195.

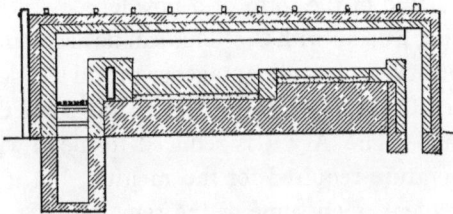

FIG. 196.

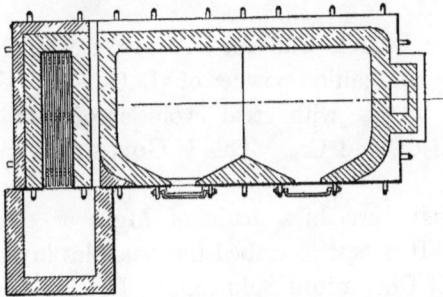

FIG 197.

FIGS. 195-7.—Roasting Furnace.

permit from 1 to 2% of neutral chromate to remain in the solution, the evaporation may take place in iron vessels. After the dehydration is almost complete, the melted $Na_2Cr_2O_7$ is poured into flat pans in which it solidifies.

4. **Reduction of $Na_2Cr_2O_7$ with S** is accomplished by melting the mixture in cast-iron kettles, which are set in masonry over a grate. $Na_2Cr_2O_7 + S = Na_2SO_4 + Cr_2O_3$. The melt is ladled out, broken up after it has become cold, and leached with water; the $Na_2SO_4$ passes into solution, while $Cr_2O_3$ remains undissolved and is separated by decanting and filtering off the solution.

## II. Reduction of the $Cr_2O_3$.

**By a Reducing Roast.** If it is not necessary for the Cr to be obtained in a fused condition, it suffices to mix the $Cr_2O_3$ with wood charcoal, or powdered coke, and heat the mixture in crucibles. Even in the regenerative gas furnace, the reduced metal is not fused, but is to be found at the bottom of the crucible in the form of a powder.

**By a Reducing Fusion with C.** This can be accomplished only in electric furnaces. It is advantageous to add a little $Al_2O_3$ to the mixture of $Cr_2O_3 + 3C$ and to use as flux some $CaF_2$ or some $AlF_3.3NaF$. The $Al_2O_3$ is reduced to metal with the Cr at the temperature required for the melting, but it then acts as reducing agent upon some of the remaining $Cr_2O_3$.

APPARATUS: Héroult or Girod furnaces. See Iron, pp. 208, 210.

**By Reducing Fusion with Al.** A mixture of $Cr_2O_3 + Al_2$ is kindled by an ignition powder of $3BaO_2 + Al_2$; the mass then continues to fuse with great evolution of heat and there is formed $Al_2O_3$ and $Cr_2$. This is GOLDSCHMIDT'S THERMITE PROCESS.

APPARATUS: crucibles made of MgO or else lined with MgO. It is best to embed the crucibles in sand.

**Electrolysis of Chromium Solutions.** The fact that chromium can be deposited electrolytically from aqueous solutions of chromous chloride was discovered by Bunsen in 1854. This scientist showed that a relatively pure metal could be produced by using high current densities (at least 70 amperes per square foot of electrode surface) and concentrated $CrCl_2$ solutions. Both conditions are difficult to maintain, for the surface of the electrode is constantly increasing by deposition of Cr and the concentration of the solution is constantly diminishing. For this reason the author in 1887 to 1900 used, instead of the $CrCl_2$ solution, a paste of $CrF_3$ crystals packed in a linen bag and suspended in a vessel of water through which $SO_2$ was constantly passed during the electrolysis to effect depolarization. The cathodes consisted of Pt foil, the anodes of C plates. The current density was gradually increased during the electrolysis, to compensate for the gain in electrode surface. The Cr grew upon the cathode foil in a crystalline condition.

By the more recent studies of G. Glaser, the limits of concentration and of current density have been established for $CrCl_2$ solutions.

ELECTROLYTE: $CrCl_2$ solutions containing 13 to 20 oz. Cr per gallon.

ANODES: rods of carbon.

CATHODES: Pt foil.

CURRENT DENSITY: 85 to 170 amperes per square foot of electrode surface. With densities of 70 amperes per square foot, the metal contains perceptible amounts of CrO, and with 8 amperes per square foot, only CrO is deposited.

TEMPERATURE: 50° C. (122° F.), at the most. At higher temperatures the Cr is not deposited in a crystalline condition, but in the form of a black powder.

## Properties of Chromium:

SPECIFIC GRAVITY: 6 to 7.

COLOR: light gray with high luster.

MECHANICAL PROPERTIES: hard and brittle.

STRUCTURE: coarsely crystalline.

MELTING-POINT: 1515° C. (2760° F.) (?).

BOILING-POINT: 2500° C. (4532° F.).

ALLOYS readily with Fe, Mn and W; difficultly soluble in most other metals. Some melted products, formerly regarded as alloys, have proved to be mixtures upon more accurate metallographical investigation In these the Cr lies finely divided in the other metal.

CHEMICAL BEHAVIOR: At low temperatures the metal is fairly stable but at high temperatures it combines energetically with most of the metalloids, particularly with C and Si. It is more readily dissolved in caustic alkali solutions than in acids. With the former, it forms salts in which the Cr is present in the anion; with the latter, salts in which the Cr exists as a bivalent or trivalent cation.

# TUNGSTEN

## Sources.

**Natural Sources:**

AN OXIDE occurs, called
Tungstite, $WO_3$.

THE SALTS, called tungstates.

Wolframite, $FeWO_4$.
Scheelite, $CaWO_4$.
$\left.\begin{array}{l}\\ \\ \end{array}\right\{$ These tungstates are more widely distributed than the oxide and accompany cassiterite (cf. Tin).

**Other Sources:**

SLAGS obtained in smelting tin ores that contain tungsten.

## (A) Ferro-Tungsten.

In the chapter on Iron, it was mentioned that tungsten is used in the manufacture of a special kind of steel, and that such steels could be prepared by the addition of an iron alloy rich in W (80 to 85% W). This alloy, called *ferro-tungsten*, is prepared by a

**Reducing Fusion** of wolframite, or scheelite, with powdered quartz and glass; these fluxes slag the bases other than iron that occur in the ores, particularly the alkaline earths. In working with wolframite it is well to remember that the ore often contains considerable amounts of hübnerite, $MnWO_4$. When scheelite is used, it is obvious that Fe must be added to the charge.

APPARATUS AND OPERATIONS are the same as with ferro chrome (p. 218).

# (B) Pure Tungsten.

Since scheelite, on account of its low Mn content, is better suited for preparing ferro-tungsten than is wolframite, which almost always contains considerable Mn, the latter is used in large quantities in the preparation of pure tungsten. In working up this ore, or tin slags likewise containing W as tungstate, there are a great many points of resemblance to the process for obtaining pure chromium.

## I. Separation of W from Fe, Mn, Ca, etc.

1. **Oxidizing Roast with Alkaline Fluxes.** Purpose: transformation of W into a soluble alkali tungstate, $Na_2WO_4$, of Fe and Mn into insoluble oxides, and of Ca into insoluble carbonate.

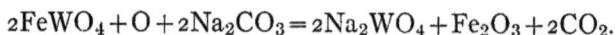

$$2FeWO_4 + O + 2Na_2CO_3 = 2Na_2WO_4 + Fe_2O_3 + 2CO_2.$$

APPARATUS AND OPERATIONS as with Chromium.

2. **Lixiviation.** Purpose: separation of the soluble tungstate from the substances that are insoluble in water. As solvent, the weak wash waters of the previously roasted product are used. The roasted product, as it comes from the furnace, is leached hot in this liquor. By thus quenching it, the roasted product is broken up where it has been sintered and put into a condition more favorable for leaching. When the solution has a concentration of 10 to $12\%$ tungstate, it is drawn off and concentrated by evaporation. Hereby certain contaminating salts, such as $Na_2SO_4$, which dissolve in water with the tungstate, separate out. When the solution is sufficiently concentrated it is cooled and brought to crystallization.

3. **Precipitation of the So-called Tungstic Acid, $WO_3$.** If the concentrated solution of tungstate is heated by introducing steam, and then hydrochloric acid is allowed to run in, or if the powdered crystals of tungstate are shaken into the

hot solution of hydrochloric acid, the $WO_3$ is deposited as a heavy, yellow powder which can be purified by decantation, filtration, and drying.

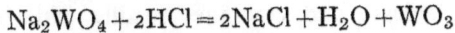

$$Na_2WO_4 + 2HCl = 2NaCl + H_2O + WO_3$$

The precipitation and first washing take place in stone vats, the last washing in filter bags.

## II. Reduction of $WO_3$.

**Reduction without Fusion.**   Mixtures of about 265 lbs. $WO_3$, 26 to 33 lbs. wood charcoal or coke, and 15 to 9 lbs. of powdered colophonium, or pitch, are placed in crucibles and heated as hot as possible in gas furnaces with blast or regenerative firing.   In this case the reduced metal is not melted.

APPARATUS AND OPERATIONS as in the reduction of Chromium (p. 222).

**Reducing Fusion.**   If the same charge as used for the above reduction is placed in an electric furnace, some difficulty will be experienced in fusing it.   The melting point of W lies very high (2800° to 2850° C.) and although a temperature of 3500° C. is reached in the electric arc, this region of high temperature is of very limited area.   Hence the electrodes must be brought as close to the metal as possible, or else the latter must be made one pole in the circuit of the arc, and by so doing it is very hard to avoid the taking up of C. Carbon is dissolved with the formation of the carbides $W_2C$ and WC.   Metal containing carbide can be fused readily in the electric arc.

## Properties of Tungsten:

SPECIFIC GRAVITY: 19.

COLOR: gray, crystalline as powder; in the fused condition it is a nearly white, lustrous metal.

MECHANICAL PROPERTIES: Very hard.

MELTING POINT: 2800° to 2850° C. (5070° to 5160° F.).

BOILING POINT: 3700° C. (7000° F.).

ALLOYS with other metals to about the same degree as Cr does.

CHEMICAL BEHAVIOR: It is oxidized by O only at high temperatures; likewise the halogens act vigorously only when the metal is hot. It is insoluble in most acids. Strongly oxidizing acids produce $WO_3$; this oxide is an acid anhydride and combines readily with basic oxides to form tungstates.

# CADMIUM

## Sources

**Natural Sources:**

THE SULPHIDE, CdS, called greenockite, is of rare occurence
and is usually accompanied by ZnS in the blendes of
Silesia and of North America.

THE CARBONATE, $CdCO_3$, is found, as a rule, only with the
smithsonites of Silesia and of North America.

**Other Sources:** The zinc dust formed in working up zinc
ores containing cadmium.

## (A) Extraction

Since there are not enough cadmium ores available to support
an independent industry, the production of cadmium forms
merely a side-issue in zinc smelters that work zinc ores containing cadmium. As will be seen under the section on Zinc
(p. 234), when such ores are roasted CdO is formed together with
ZnO and it is the former oxide that is first reduced to metal.
The Cd is found, therefore, partly as metal, and partly as carbonate
and oxide, together with large amounts of Zn and ZnO (70 to
80%) in the first zinc dust that is obtained. The preparation of
Cd from this raw material consists in a repeated

**Reduction with Fractional Distillation,** first in clay and finally
in iron retorts. The method of working is the same as for
the reduction of ZnO, except that the temperature is kept
lower. For details, see Zinc.

# (B) Refining

The crude cadmium, containing more or less zinc, can be purified further by repeating the above-mentioned

**Reduction** with fractional distillation until finally the Zn content is brought very low. It can also be freed from Zn by

**Electrolysis.**

ANODES: crude Cd containing Zn.

CATHODES: pure Cd.

ELECTROLYTE: $CdCl_2$ or $CdSO_4$.

CURRENT DENSITY: 6 to 15 amperes per square foot.

E.M.F.: 2.8 to 3.5 volts.

**Properties of Cadmium:**

SPECIFIC GRAVITY: 8.6 to 8.7.

COLOR: white, strongly lustrous.

MECHANICAL PROPERTIES: soft, ductile.

STRUCTURE: It shows a tendency to form dendrites on the surface, otherwise granular. The crystal grains grow considerably upon long-continued heating.

ALLOYS with most other metals. It forms some alloys with remarkably low melting-points; e.g., 4 pts. Bi, 1 pt. Sn, 2 pts. Pb, 1 pt. Cd = Wood's Metal. 15 pts. Bi, 4 pts. Sn, 8 pts. Pb, 3 pts. Cd = Lipowitz Metal. The former alloy melts at 71° C. (160° F.) and the latter at 60° C. (140°F.).

CHEMICAL BEHAVIOR: At low temperatures it is stable in the air but unites readily with the halogens. At high temperatures it burns readily in O or S. It is readily soluble in HCl, $H_2SO_4$, and $HNO_3$, also in alkali hydroxides. E.M.F. toward H = +0.420 volt.

# ZINC

## Sources

**Natural Sources:**

ZINC BLENDE, or sphalerite, ZnS. Gangue and accompanying minerals are

1. Galena in Germany in the ore deposits at Ems and the Upper Harz, also in Bohemia and Hungary, and in Australia.

2. Dolomite, limonite, and also smithsonite, often in clay, in Rhineland, Westphalia, Belgium, Upper Silesia, Northern Spain, Algiers, and North America.

3. In gneiss with pyrites in Sweden.

SMITHSONITE* or Zinc Spar, $ZnCO_3$. Gangue: often with sphalerite and galena, always with zinc silicates in limestone, dolomite and also limonite in Rhineland, Belgium, North Spain, England, North America.

CALAMINE, $H_2Zn_2SiO_5$.

WILLEMITE, $Zn_2SiO_4$. The gangue and accompanying minerals are the same as with smithsonite.

RED ZINC ORE or zincite ZnO. Gangue; the same as with franklinite.

FRANKLINITE: (Zn, Fe, Mn) O (Fe$_2$, Al$_2$, Mn$_2$) O$_3$. It is found in New Jersey near the village of Franklin (whence the name) associated with zincite, willemite, rhodonite, and tephroite, etc.

**Other Sources:**

ZINC DUST: the first condensation product in the reduction of zinc with sometimes as much as 90% Zn.

---

* Sometimes called calamine, but incorrectly

FLUE DUST from smelting furnaces.

FURNACE CALAMINE, the deposit in shaft furnaces in which ores containing Zn are smelted.

DROSS, waste metal obtained in the refining of zinciferous metals, and waste from foundries, plating works, etc.

ZINC SKIMMINGS, alloys of Ag, Pb, Cu and Zn obtained in the removal of Zn and Ag from lead bullion.

# (A) Extraction of Crude Zinc

## Roast Reduction Work:

1. **The Roasting** effects the transformation of ZnS and other sulphides, as well as of $ZnCO_3$ and other carbonates, into oxide. Any water held mechanically or chemically is also driven out.

The principal reactions that take place are

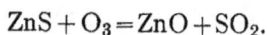

$$ZnCO_3 = ZnO + CO_2$$

$$ZnS + O_3 = ZnO + SO_2.$$

As compared with the other sulphides, sphalerite is difficult to roast; basic sulphates as well as oxide are formed and the former are hard to decompose. Both the oxide and the basic sulphates form a coating over the sulphide and prevent the access of air. A complete roasting, which is necessary on account of the injurious action of sulphides upon the retorts, can be accomplished only with the application of external heat.

ROASTING APPARATUS:

HEAPS AND OPEN KILNS for drying, and rarely for calcining, material containing carbonates.

SHAFT FURNACES for burning carbonate material.

REVERBERATORY FURNACES with fixed hearths, worked by hand or in some cases provided with mechanical stirrers. Recently for roasting blende that contains pyrite, reverberatory furnaces with revolving hearths have been sometimes used.

MUFFLE FURNACES, first introduced by Liebig and Eichhorn by arranging heating flues in the fine pyrites burner of the Maletra-Schaffner System, and then, with success in zinc smelters, by Hasenclever. In furnaces of this type, four men (2 for each 12 hours) can roast from 4 to 4.5 tons of

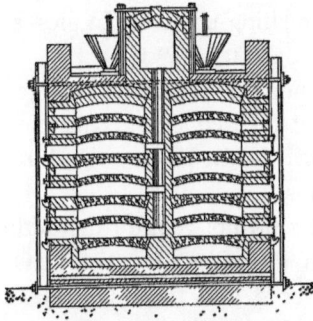

FIG. 198.

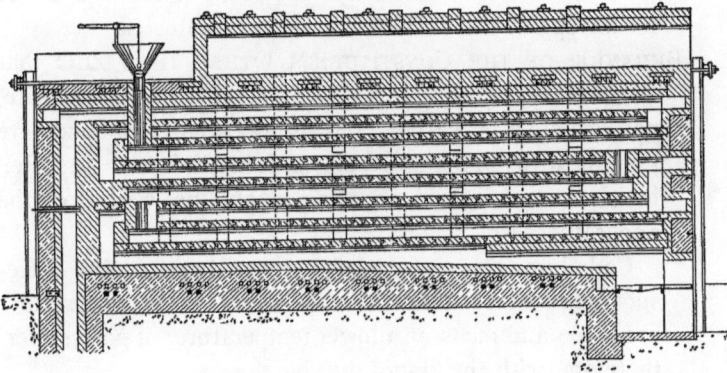

FIG 199.
Hasenclever Furnace.

ore in a day. From 300 to 450 lbs. of coal are required per ton of ore.

2. **Reduction of the Roasted Product.** To carry out the work to the best advantage, the properties of zinc oxide and of zinc necessitate the removal of the metal from the reduction apparatus in the form of vapor and the condensa-

tion of the vapor in receivers; in other words a distillation is combined with a reduction. Although the reduction of zinc oxide by carbon will take place at a distinct red heat, still the reaction under these conditions is so slow and incomplete that to obtain satisfactory results the reduction temperature of the zinc oxide must be kept between $1000°$ and $1300°$ C.; the melting-point of Zn lies at $415°$ C. and the boiling-point at $930°$ to $950°$ C.

FLUX: An abundance of carbon to prevent the formation of $CO_2$ during the reduction; this gas will oxidize Zn to ZnO at red-heat in the receivers and results in the formation of zinc dust.

Zinc silicates do not require the addition of any flux to combine with the silica, for the ZnO contained in the silicate is reduced and the $SiO_2$ remains behind as such or in the form of acid silicates, and tends to make the residue fusible; the more difficultly-fusible the residue the better for the apparatus in which the work is conducted.

BEHAVIOR OF THE CONSTITUENTS OTHER THAN ZnO that are present in the charge: Zinc sulphide is, to be sure, not reducible by carbon, but usually iron compounds are present with which it reacts to form FeS and Zn; FeS has a very injurious action upon the walls of the apparatus.

Cadmium compounds behave similarly to the corresponding zinc compounds except that CdO is more readily reducible and melts at a lower temperature; it passes over, therefore, with the first of the zinc dust.

Of the iron and manganese compounds, $Fe_2O_3$ and $Mn_2O_3$ are reduced to FeO, Fe, and MnO. FeO and MnO unite with silicates to form readily-fusible slags and these are dangerous for the retorts. FeS is unacted upon, but melts and injures the walls.

Lead oxide in the free state is easily reduced and is volatilized to some extent with the Zn; combined with silica, it forms fusible silicates which injure the walls.

Free silica is useful, since the melting-point of the residue increases in proportion to the $SiO_2$ content.

Under the different modes of working that prevail in different countries, and in connection with different mines, the usual apparatus and processes followed may be classified into three groups: the Silesian, the Belgian, and the Rhenish methods.

In all cases, muffles, or tube-shaped retorts, closed at one end, are used as reduction vessels. The retorts are either oval, semi-oval, or cylindrical in form. The size and shape of the retort are regulated by the following circumstances.

ZINC CONTENT: the greater the zinc content, the smaller the retort.

REDUCIBILITY: the harder it is to reduce the ore, the smaller the retort.

FUEL: the influence of the fuel, now that gas furnaces are used almost universally, steps into the background. Formerly, furnaces with a great many small retorts required a fuel that would give a long flame, and furnaces with large retorts could get along with a short-flame fuel.

These factors have led to the development of the following systems of retorts with furnaces suitable for them:

In Silesia, where the ores are poor and where to some extent coal giving short flame is available, the work is carried out in large muffles. The Silesian muffles have the following inside dimensions: height 26 in., breadth 6 to 8 in.; length 27 to 85 in.; thickness of walls, $\frac{3}{4}$ in. in front and $2\frac{1}{2}$ in. in back.

THE CONDENSERS for Silesian muffles are metal tubes, metal boxes, and also clay tubes, conical or bellied in form.

CAPACITY: up to 220 lbs. ore, together with about 40% as much of reduction carbon.

ARRANGEMENT IN THE HEATING CHAMBERS: in a row, as many as 36 pieces side by side; two heating chambers lie with the vertical back walls opposite one another so that one furnace may hold as many as 72 muffles.

FIRING: usually regenerative.

CONSUMPTION OF FUEL: one Silesian furnace with 60 muffles, each of 220 lbs. capacity, will work up 6 tons of ore in 24 hours, using thereby 6.1 tons of heating carbon and 2.4 tons of reducing carbon.

YIELD: 1 ton of zinc; hence 6.1 tons of heating carbon and 2.4 tons of reducing carbon are used per ton of zinc.

IN BELGIUM where the ores are rich and long-flame coal is available, small retorts are used which are either

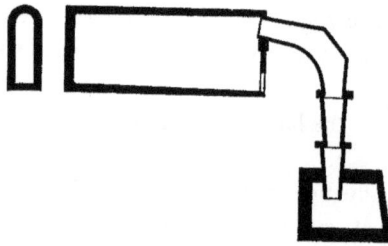

FIG. 200.—Silesian.

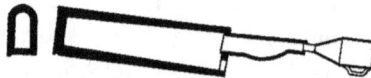

FIG. 201.—Rhenish.

FIG. 202.—Belgian.
Types of Muffles. Scale 1 : 50.

cylindrical tubes 6 to 10 in. in diameter, or oval tubes $6\frac{1}{4}$ to 7 in. wide and $7\frac{7}{8}$ to 11 in. high; these retorts are 40 to 57 in. long and the walls are $\frac{3}{4}$ in. thick in front and $1\frac{1}{2}$ in. thick in the rear.

CONDENSERS FOR THE BELGIAN RETORTS: conical clay tubes 16 in. long, and with a diameter of 3 in. in front and 6 in. in back. The wide end is shoved into the retort and the narrow end is fitted into a sheet-iron condenser, or nozzle, which serves to catch the zinc dust.

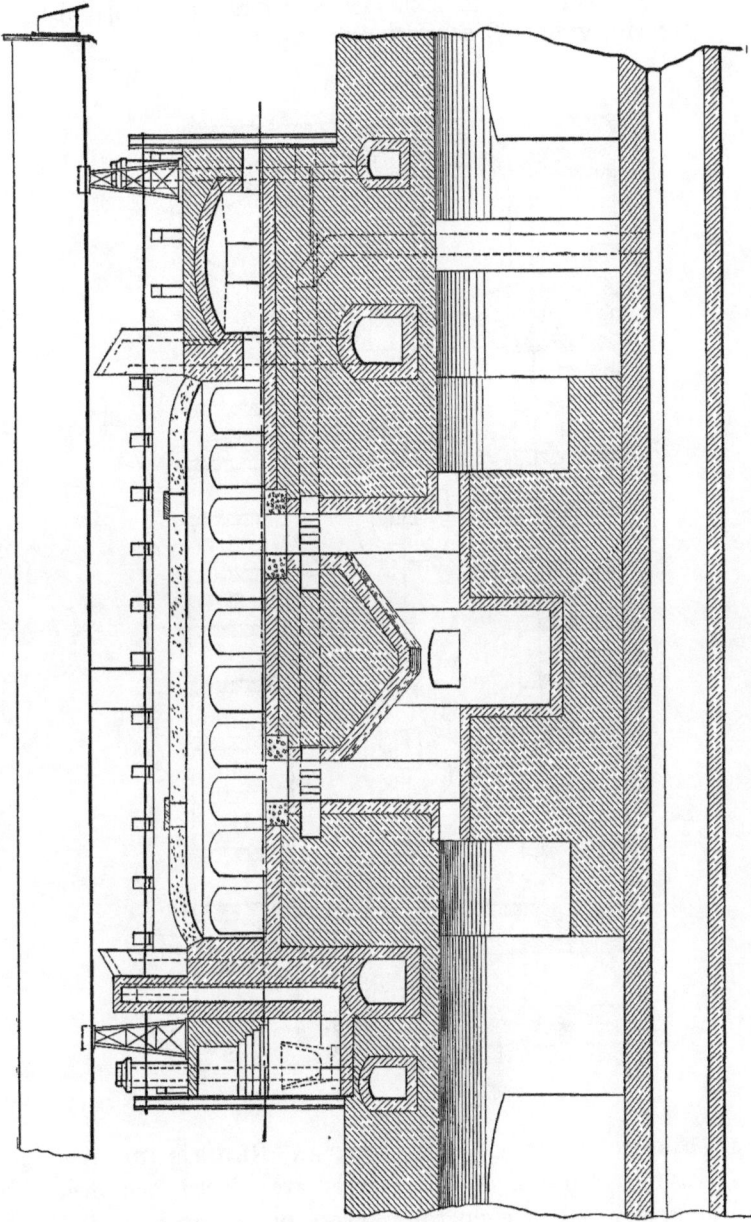

FIG. 203.—Section of Silesian Zinc Furnace, Dombrowa (Schrubko).   Scale 1 : 100.

CAPACITY OF THE BELGIAN RETORTS: about 44 lbs. ore with 40 to 60% as much reducing carbon.

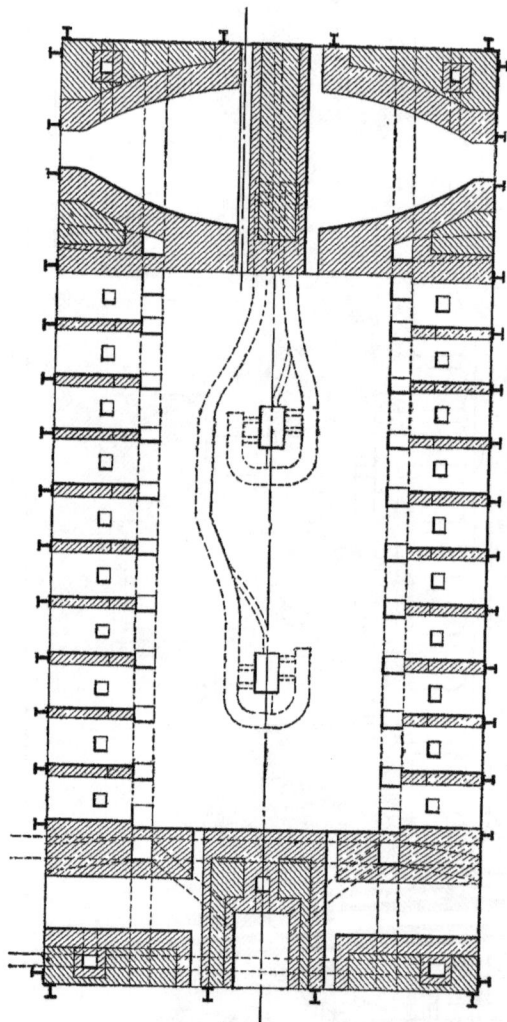

FIG. 204.—Plan of Silesian Zinc Furnace, Dombrowa (Schrubko). Scale 1 : 100.

ARRANGEMENT OF THE BELGIAN RETORTS IN THE HEATING CHAMBERS: the retorts are placed one over another in six or eight rows; every two or three vertical series open into a common niche in which the condensers lie. The back end of the tube rests against a supporting wall, and the front end in the wall of the niche; the rest lies free. In one furnace-block, with, as a rule, the back walls of two adjacent heating chambers lying against one another, there are from 56 to 400 of these tubes or retorts.

FIRING: Siemens' regenerative or recuperative firing of different systems; direct coal firing is seldom used in modern plants; in the United States natural gas is also employed as fuel.

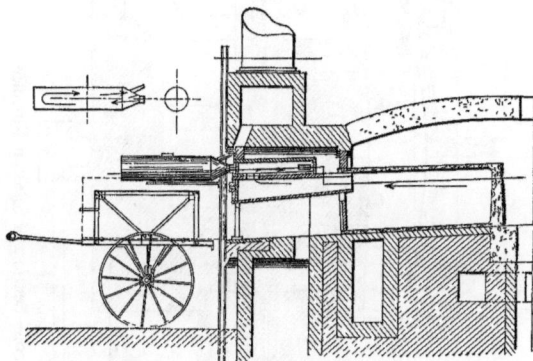

FIG. 205.—Part of Silesian Furnace

CONSUMPTION OF FUEL: furnaces with 56 or 70 retorts require respectively for 1.0 and 1.35 to 1.44 tons of ore, 1.8 and 2.5 to 2.55 tons of fuel.

YIELD: One ton of zinc produced required 4.5 to 5.5 tons of fuel, and 1.3 to 1.9 tons of reducing carbon.

IN ZINC SMELTERS OF THE RHENISH PROVINCES the attempt has been made to combine the advantages of large muffles with the possibility of arranging them as is customary with the smaller retorts. Semi-oval retorts are used, varying in size between the Silesian muffles and Belgian retorts, and so chosen that, with the full burden, they can be placed free in the heating chamber according to the Belgian method.

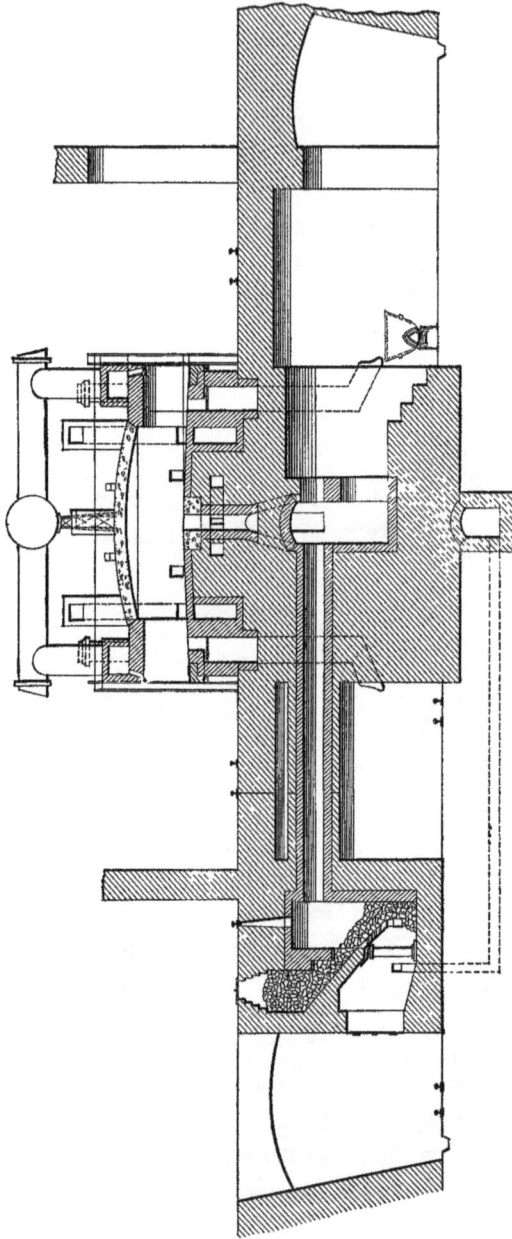

FIG. 206.—Section of Silesian Furnace.

The dimensions of the retorts in common use in the Rhenish provinces are: breadth up to $6\frac{1}{4}$ in., height up to $11\frac{3}{4}$ in., length up to 55 in.

CONDENSERS FOR THE RHENISH RETORTS: bulging clay tubes open at both ends, of which the wider end is

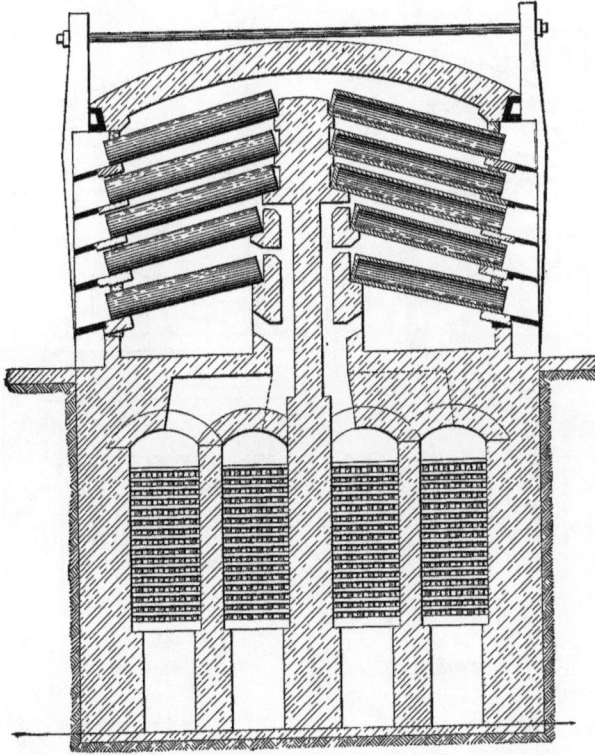

FIG. 207.—Belgian Furnace with Siemens, Regenerative Chambers.

inserted into the end of the retort and, at the beginning of the work, the front ends are stuck into the sheet iron prolongs, so as to catch the zinc dust.

CAPACITY OF THE RHENISH RETORTS: 55 to 75 lbs. of ore with 40 to 50% as much reducing carbon.

ARRANGEMENT OF THE RETORTS IN THE HEATING CHAMBERS: the retorts rest above one another in three

horizontal layers; every two vertical columns open into
a front niche in which are the receivers. The retorts
lie as in the Belgian type. The furnace block may contain 40 to 80 retorts in three series, i.e., 120 to 240 retorts
in all.

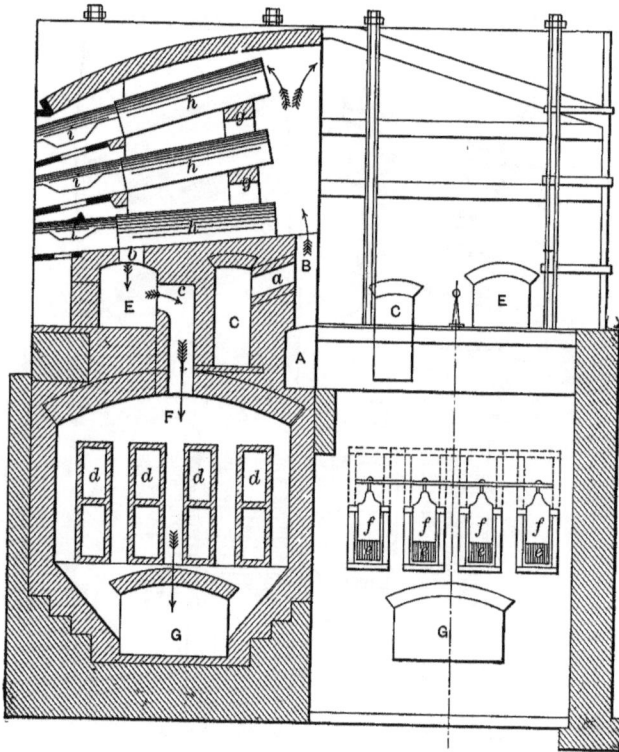

FIG. 208.—Rhenish Furnace with Recuperative System.

FIRING: Siemens' regenerative firing is less used than
the so-called *recuperative firing*.

CONSUMPTION OF FUEL: a large furnace of this type
with 240 retorts and recuperative firing requires 8.5 to
9.2 tons of fuel and 3.2 tons of reducing carbon in 24 hours
for operating with 8 tons of ore with 52 to 54% of zinc.

YIELD: 2.5 tons of fuel and 1 ton of reducing carbon are required to produce one ton of zinc.

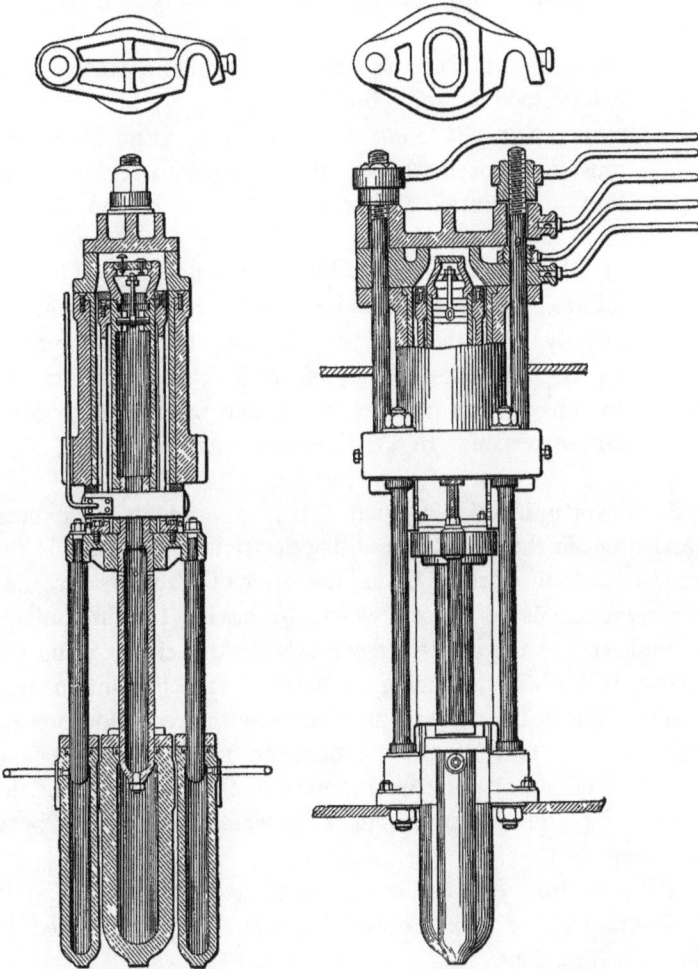

FIG. 209.—Retort Press (C. Mehler, Aachen).   Scale 1 : 40.

THE MAKING OF THE RETORTS is carried out as a rule at the smelter.  The work consists of the following operations: BURNING of a part of the clay that is to be used in making the retorts.

GRINDING.

PULVERIZING the burnt clay.

MIXING the powdered clay with 50 to 100% as much fresh clay.

STAMPING AND PRESSING the mixture into retorts after it has stood for some time.

DRYING AND BURNING THE RETORTS: Enough retorts must be kept in the burning furnace to replace those that are injured during the reduction of the zinc so as to avoid the danger of injury that would be likely to take place in the cooling and reheating. The hot retorts, therefore, are introduced from this furnace directly into the retort-furnace as they are required.

LIFE OF MUFFLES. The retorts last, according to the amount of work done in the smelter, from 14 to 30 days; this corresponds to a daily replacement of 3 to 7%.

It is worth mentioning that many experiments have been carried out in the attempt to utilize electricity for heating in the roast-reduction work. Thus the Cowles Brothers in 1885 proposed the use of retorts which, by having carbon contacts at both ends, were connected with an electric circuit using the charge as the heat-producing resistance. The constant changes taking place in the body of the charge by the reduction process and the fact that as the temperature rises the retort and the walls take part more and more in the conduction of the current, have given rise to difficulties which have not yet been overcome.

**The Precipitation Process,** consisting of the smelting of sulphide ores with iron or fluxes containing iron, has been tested in different directions.

In the blast-furnace, the possibility of such a process is altogether out of the question, because the reaction gases can never be kept absolutely free from $SO_2$ and $CO_2$. Thus the Zn would never be obtained as a fused metal.

Experiments in which the ZnS was exposed to the action of

molten Fe in a rotary, closed converter, also proved fruitless.
The requisite mixing of the ZnS and Fe, which is necessary
for the rapid fulfilment of the desired reaction, could not be
accomplished by revolving the drum.

Apparently the problem has been solved, however, by heating
a mixture of finely-divided ZnS with equally fine Fe in an electric furnace in which a bath of FeS is formed by the reaction

$$ZnS + Fe = FeS + Zn,$$

and this kept as a heating resistance. The FeS holds Fe as
well as ZnS in solution, so that it serves to maintain the desired
state of subdivision, and the reaction between the Fe and
ZnS continues to take place smoothly.

**The Direct Electrolytic Process** for working up crude zinc from
ores by making the latter the anode in an aqueous solution
of zinc salt, has never met with success.

# (B) Zinc Refining

**By Liquating** in reverberatory furnaces, in which the bottom of the
hearth falls away from the fire-bridge toward the opposite
wall, it is customary to purify the crude zinc, obtained from
the Roast-Reduction Process, which usually contains Pb. From
the fused zinc a specifically heavier alloy of Pb or Fe with Zn
separates out at the bottom and is removed from time to time
from the deepest part of the hearth. On top of the metal bath
float the more infusible impurities which were mechanically
enclosed by the crude zinc; these are skimmed off. The zinc
itself is usually dipped out with iron ladles, poured into plates,
and rolled at a suitable temperature.

**Distillation** is employed with alloys rich in lead, containing
precious metals (skimmings rich in Zn from the Zn-desilveriza-
tion).

APPARATUS:
Crucibles in air-furnaces; this is the oldest process for
small works.

Bottle-shaped distilling vessels in tilting air-furnaces. (Figs. 210 to 212.)

Bottle-shaped or tube-shaped distilling apparatus made of fire-clay with graphite lining, heated in reverberatory furnaces. (Figs. 213 and 214.)

**Electrolytic Separation** of Zn alloys has not met with success in practice (e.g., for working up skimmings rich in Zn).

**Lixiviation and Electrolytic Deposition** has been attempted at some places without meeting with commercial success, but a practical process has been worked out at the works of Brunner, Mond & Co., at Norwich, England, using the patents of Höpfner. It consists of the following operations:

**Chloridizing Roast** of Zn ores, or pyritic residues containing Zn, in reverberatory furnaces.

**Lixiviation** with water.

**Purification** of the liquor by

CRYSTALLIZING out of the sulphate or decomposing it with $CaCl_2$.

PRECIPITATION of Fe and Mn by means of $Ca(OCl)_2$.

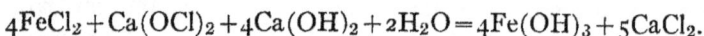

$$4FeCl_2 + Ca(OCl)_2 + 4Ca(OH)_2 + 2H_2O = 4Fe(OH)_3 + 5CaCl_2.$$

PRECIPITATION of the more electro-positive metals with zinc dust.

**Electrolysis of the Liquor.**

ANODES: carbon plates.

CATHODES: rotating Zn plates.

ELECTROLYTE: $ZnCl_2$ solution with 0.08 to 0.12% free HCl.

CURRENT DENSITY: 10 amperes per square foot.

E.M.F.: 3.3 to 3.8 volts.

THE APPARATUS is very complicated, because the anode plates must be placed in special cells, and each be provided with piping for carrying away the chlorine set free during electrolysis.

Although the practical results have never been encouraging, the sulphatizing roast of Zn ores would seem more advantageous

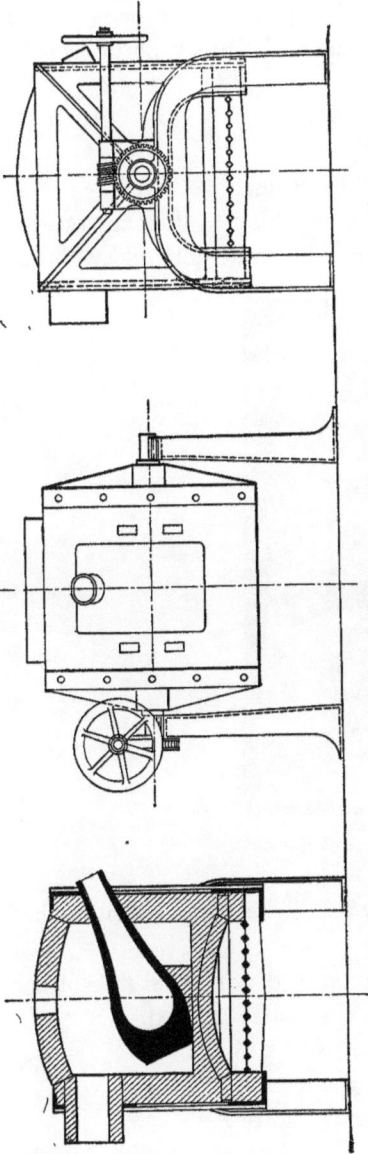

FIG. 212.

FIG. 211.
Faber Du Faur's Furnace.    Scale 1 : 50.

FIG. 210.

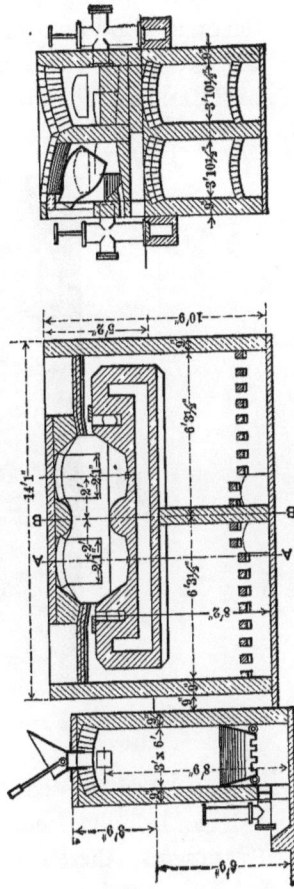

FIG. 214.

FIG 213.—Furnace at the Port-Pirie Smelter, Broken Hill, Australia

as the sulphate liquors can be electrolyzed without the use of diaphragms. Moreover, useful by-products can be produced at the anodes and in this way the E.M.F. may be utilized to the best advantage.

## Properties of Zinc:

SPECIFIC GRAVITY: 6.9 to 7.2.

COLOR: bluish-white with metallic luster.

MECHANICAL PROPERTIES: very ductile at temperatures between 100° and 150° C., can be rolled into sheets, ham-

FIG. 215.— Structure of Bar Zinc.

mered and pressed; at 200° C. it becomes so brittle that it can be pulverized.

STRUCTURE: there is a formation of dendrites on the surface of the metal, but the interior is coarse grained.

MELTING-POINT: 419° C. (786° F.)

BOILING-POINT: 940° C. (1724° F.)

ELECTRICAL CONDUCTIVITY: 0.27 that of Ag.

ALLOYS with most of the metals. The alloys with Au, Ag and Pb (zinc desilverization) have been mentioned.

Furthermore there are many alloys in practical use, such as those with Cu (brass), with Cu and Sn (bronze), with Cu and Ni (German silver), and with Sn and Sb (bearing metal).

CHEMICAL BEHAVIOR: Zinc is oxidized superficially when exposed to the atmosphere, but the layer of oxide is thin and protects the metal within so that zinc objects made of thin foil will keep for a very long time. It is attacked by the halogens at ordinary temperature and by the other metalloids at higher temperatures. The metal will reduce $H_2O$ at a moderate red heat. On account of its high electrolytic solution tension (E.M.F. toward $H = +0.770$ volts), it is easily soluble in HCl, $H_2SO_4$, $HNO_3$, and in caustic alkali solutions. With the acids it forms salts in which the metal is present as a bivalent cation; with the alkalies it forms zincates in which the Zn is in the anion.

# MANGANESE

## Sources

**Natural Sources:**

As Oxides and Hydrated Oxides in the following ores:
Braunite, the normal manganic oxide, $Mn_2O_3$.

Hausmannite, mangano-manganic oxide, $Mn_3O_4$ (this can be regarded as $Mn_2MnO_4$, manganous manganite).

Pyrolusite, black oxide of manganese, manganese peroxide, or dioxide, $MnO_2$, the most widely distributed ore of manganese.

Manganite, $Mn_2O_2(OH)_2$.

Salts:

Rhodocrosite, or manganese spar, $MnCO_3$. This is an important ore used in the preparation of Mn-Fe alloys.

Rhodonite, $MnSiO_3$.

**Other Sources:**

Slags: particularly from pig iron mixers and the preparation of alloys rich in Mn.

Metal Waste, ferro-manganese.

## (A) Manganese Alloys

The large amounts of Mn used in the metallurgy of iron, together with the fact that it is difficult to prepare from pure ores by the relatively inexpensive smelting processes a pure metal that is stable in the air, make the most important work of manganese smelting the preparation of

**Iron-manganese Alloys:**

**Spiegeleisen,** a name given to alloys with between 5 to 20% Mn.

**Ferromanganese,** which includes the richer alloys with 30 to 80% Mn. The preparation of either of these alloys takes place by a reducing smelt, such as was described under Iron, of iron ores containing manganese, or of manganese and iron ores together in a blast furnace. The working conditions are changed only in so far as is necessitated by the fact that as the Mn content of the required alloy increases, the temperature of the furnace and the content of bases in the slag (particularly of MnO) must be raised. The latter contains from 8 to 30% MnO. High slag losses of the Mn are unavoidable, because if a poorer slag were formed, the primarily reduced Mn would react, as it always does, with the oxides of iron in the ore. Furthermore, large losses of Mn occur through volatilization.

# (B)  Manganese Compounds

Among the manganese compounds which come into consideration for metallurgical purposes, or which can be prepared in smelters, the only ones to be mentioned are,

**Manganese Silicides,** or alloys of manganese silicide. One of these silicides has the composition $Mn_2Si$ and contains approximately 20% Mn. There is only one commercial method for preparing silicides, and this consists of a reducing smelt of manganese oxides with sand in an electric furnace. Since, however, most Mn ores contain enough Fe to make the melted silicide unsuitable for some purposes, such ores are preferably smelted first for a ferromanganese and a slag free from Fe which, with the addition of sand, will yield a silicide practically free from Fe with from 20 to 22% Mn.

# ˙ (C) Pure Manganese

Pure manganese cannot be obtained by smelting the purest manganese oxides with C. Even when the amount of C in the charge is most carefully restricted, the metal obtained will contain $Mn_3C$ and as this compound is decomposed by water at ordinary temperatures, it is not stable in moist air. A fairly stable metal, low in carbon, but one which is not perfectly pure, can be obtained by the

Thermite Process of Goldschmidt. $Mn_2O_3$ is mixed with $Al_2$ and the reaction started by means of ignition powder ($3BaO_2 + Al_2$). The heat of reaction is so great that the reduced Mn is entirely liquefied and the slag of $Al_2O_3$ separates out of itself.

## Properties of Manganese:

SPECIFIC GRAVITY: 7 to 8.

COLOR: white, lustrous.

MECHANICAL PROPERTIES: very hard and brittle.

FRACTURE: smooth, almost vitreous.

MELTING-POINT: 1233° C.(2251° F.)

VOLATILIZES considerably even at the melting temperature.

BOILING-POINT: 2200° C.(4000° F.)

ALLOYS with Fe, and very readily with Cu and the precious metals.

CHEMICAL BEHAVIOR: Mn is one of the metals that enters most readily into chemical reactions. It has a great affinity toward almost all of the metalloids and possesses a fairly high electrolytic solution tension, both in aqueous solutions of acids (+1.075 volts toward H), as well as in the melts that come into consideration in metallurgical work (a good desulphurizing agent for pig iron).

Mn forms a series of six oxides with corresponding salts. Those oxides with the least O possess basic properties, whereas those containing more O are acid anhydrides.

# ALUMINIUM

## Sources

**Natural Sources:**

OXIDES AND HYDRATED OXIDES.

CORUNDUM, pure $Al_2O_3$. The ruby and sapphire are varieties of corundum.

EMERY, impure $Al_2O_3$.

BAUXITE, hydrargillite, $Al(OH)_3$, together with $Fe(OH)_3$.

SALTS:

CRYOLITE, $AlF_3.3NaF$.

ALUM, $Al_2(SO_4)_3$, crystallized with a sulphate of a univalent metal.

SILICATES, particularly the feldspars and the weathering products of the feldspars, e.g. $CaO.Al_2O_3.2SiO_2$.

## Extraction

It is impossible to obtain a crude Al and to refine it by the smelting processes such as have been described for other metals, because the metal is too electro-positive. The only ores that come into consideration, therefore, are those which occur in a relatively pure state, or which can be purified without difficulty. Cryolite belongs to the former class and hydrargillite, or bauxite, to the latter. In the present method of preparing aluminium, cryolite is used as solvent for bauxite.

**I. Purification of Crude $Al(OH)_3$.** This is accomplished by the following operations:

1. **Transformation into $Na_3AlO_3$.** According to the old-fashioned way of roasting with soda

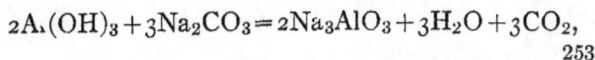

$$2Al(OH)_3 + 3Na_2CO_3 = 2Na_3AlO_3 + 3H_2O + 3CO_2,$$

or by dissolving in caustic soda, carrying out the reaction in iron kettles under a pressure of 5 to 7 atmospheres.

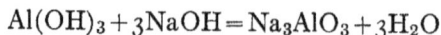

$$Al(OH)_3 + 3NaOH = Na_3AlO_3 + 3H_2O$$

In the last process it is necessary to roast the ore slightly before treating with the caustic soda, so that the FeO will be converted into $Fe_2O_3$.

2. **Leaching of the Sodium Aluminate in the old process;** in both cases there is filtration, leaving $Na_3AlO_3$ in solution, and a residue of $Fe_2O_3$ and the gangue of the ore.

3. **Decomposition of the Aluminate by $CO_2$.**

$$2Na_3AlO_3 + 3H_2O + 3CO_2 = 2Al(OH)_3 + 3Na_2CO_3.$$

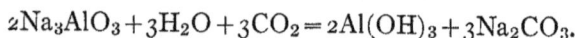

The $Al(OH)_3$ separates out from the solution and is obtained by filtration, washing and drying.

II. **Electrolysis of $Al_2O_3$ in a Molten Bath.** The Héroult process is based upon the electrolytic decomposition of $Al_2O_3$ held in solution by melted $AlF_3.3NaF$, using, at the same time, the electric current for fusing the electrolyte and deposited metal.

The working conditions as carried out to-day are:

ELECTROLYZING VESSELS: wrought iron vats with rectangular cross-section, 3 to 4.5 ft. long, 20 to 30 in. wide. The vats are lined with carbon at the bottom and on the sides with cryolite. Both of these linings are kept cold enough by air, which is made to play about the vessels by artificial draft, so that no O enters into the metal from the bottom, and none of the side lining melts.

ELECTROLYTE: $Al_2O_3$ dissolved in fused $AlF_3.3NaF$.

ANODES: Carbon blocks introduced into the vat from above.

CATHODES: at the start, the carbon bottom lining of the vessel; later, the metallic Al that separates out there.

TEMPERATURE: 750° C. (1382° F.).

CURRENT DENSITY: 700 amperes per square foot of cathode surface (the horizontal cross-section of the bath).

BATH POTENTIAL: The loss in E.M.F. corresponds on an average to 7.5 volts per furnace.

CONSUMPTION OF POWER: 1400 electric horse power per ton (2200 lbs.) of metal produced in 24 hours.

PROCEDURE AND REACTIONS DURING THE ELECTROLYSIS. To start the process, the upper electrodes are lowered until they form an electric arc with the carbon bottom, then $AlF_3$ is charged, and the electric arc is maintained until the amount of fused cryolite is sufficient for the upper electrodes to dip into it, then the addition of cryolite is

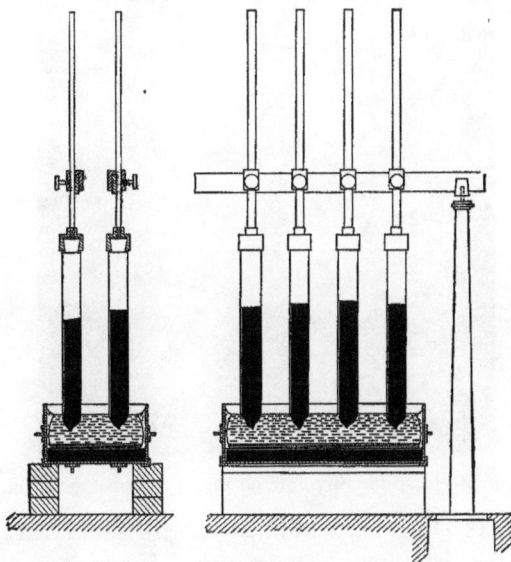

FIG. 216 —Electric Furnace of Héroult Type.

continued until the melt is 8 to 12 in. deep in the vat. Now, as Al separates out during the electrolysis, it is replaced in the bath by a corresponding amount of $Al_2O_3$, carefully avoiding a large excess of the latter. The metallic Al separates out at the bottom cathode and O is evolved at the anodes which dip into the melt, and this O combines with the C to form $CO_2$. The Al is removed from time to time by tapping or ladling.

Although quite a number of other electrolytic processes have proved suitable for the production of aluminium on a large scale, they have not shown so great economy as the Héroult process. They possess a historical interest, however, and a few will be mentioned as representing the gradual development of the Al industry.

BUNSEN in 1854 deposited the first Al from a fused compound, $AlCl_3$.

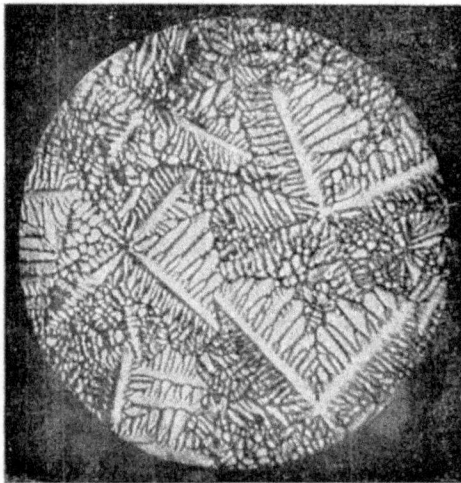

FIG. 217 —Dendritic Structure on Lower Side of Cast Ingot.

ST. CLAIRE DEVILLE, a little later in the same year, proposed that the deposited Al be replaced in the bath by $Al_2O_3$, but soon after gave up electrolysis and proposed the decomposition of $AlCl_3.NaCl$ by Na, the method that was used for a long time in the technical production of aluminium.

ROSE, in 1855, recommended the decomposition of cryolite by Na.

BEKETOFF, in 1865, the decomposition of cryolite by Mg.

GRABAU, in 1887, started with $Al_2(SO_4)_3$ and discovered the following interesting reactions:

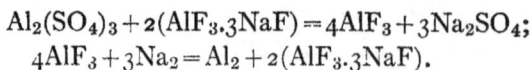

$$Al_2(SO_4)_3 + 2(AlF_3.3NaF) = 4AlF_3 + 3Na_2SO_4;$$
$$4AlF_3 + 3Na_2 = Al_2 + 2(AlF_3.3NaF).$$

COWLES BROTHERS, in 1884, carried through the reducing smelting of $Al_2O_3$ with C in the electric furnace but could not obtain pure Al; they worked, therefore, for the production of Al alloys with Cu and Fe: aluminium bronze and ferro-aluminium.

## Properties of Aluminium:

SPECIFIC GRAVITY: 2.53.

COLOR: white with high luster.

MECHANICAL PROPERTIES: at 100° to 150° C. (212° to 302° F.), it can be forged and rolled satisfactorily; at 500° C.

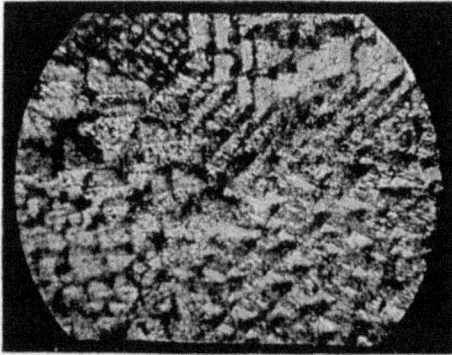

FIG. 218.—Large-grained Inner Surface of a Partly Solidified Block from which the Still-molten Metal has been Withdrawn.

(932° F.) it can be compressed easily; at 530° C. (986° F.) it becomes so friable that it can be pulverized.

STRUCTURE: according to the way it is poured, it is either dendritic or grained.

MELTING-POINT: 687° C. (1270° F.).

ALLOYS with most other metals. In the experiments carried out in the attempt to prepare Al, and as regards technical importance, the alloys with the following metals have been studied: Cu, Ni, Co, Fe, Sb, Sn, Zn, Mg.

CHEMICAL BEHAVIOR. At ordinary temperatures Al is very resistant to the constituents of the atmosphere; even at 700° to 800° C. it oxidizes only slowly, but with the development of a great deal of heat $1$ kg. $= 6274$ Cal. ($1$ lb. $=$ $11,293$ B.T.U.). It combines with halogens even at low temperatures and readily with the other metalloids; even with N, at high temperatures. In spite of its high solution tension ($+1.276$ toward H) it dissolves comparatively quickly only in HCl and NaOH; in $H_2SO_4$ it dissolves much less readily, and in $HNO_3$ very slowly. Al, however, is a very energetic reducing agent and precipitant for metal oxides and other compounds of metals, both in the solid, molten and dissolved condition. Particularly in the decomposition or oxides rich in O, so much heat is evolved that difficultly-fusible metals, like Cr and the slag of $Al_2O_3$, are melted or even volatilized (cf. Chromium, page 222). In the compounds that are of industrial importance, Al is either present as a trivalent cation (of the type $Al_2O_3$) or, when combined with strongly basic oxides, in the anion (forming aluminates, of the type $H_3AlO_3$).

# INDEX

259

# SHORT-TITLE CATALOGUE

OF THE

## PUBLICATIONS

OF

## JOHN WILEY & SONS

### NEW YORK

#### LONDON: CHAPMAN & HALL, LIMITED

ARRANGED UNDER SUBJECTS

Descriptive circulars sent on application    Books marked with an asterisk (*) are sold at *net* prices only.    All books are bound in cloth unless otherwise stated.

## AGRICULTURE—HORTICULTURE—FORESTRY.

Armsby's Principles of Animal Nutrition.. ...................... ..8vo, $4 00
* Bowman's Forest Physiography   . . . . . . . . .................8vo,   5 00
Budd and Hansen's American Horticultural Manual.
    Part I   Propagation, Culture, and Improvement.............. 12mo,   1 50
    Part II. Systematic Pomology .. ....................... 12mo,   1 50
Elliott's Engineering for Land Drainage... ....... . ............12mo,   2 00
    Practical Farm Drainage.  (Second Edition, Rewritten )... ... 12mo,   1 50
Fuller's Water Supplies for the Farm.  (In Press.)
Graves's Forest Mensuration . . .... . ......... ............. ... 8vo,   4 00
    * Principles of Handling Woodlands  ...  ............ Large 12mo,   1 50
Green's Principles of American Forestry.. ......................12mo,   1 50
Grotenfelt's Principles of Modern Dairy Practice    (Woll.).... ......12mo,   2 00
Hawley and Hawes's Practical Forestry for New England.  (In Press.)
* Herrick's Denatured or Industrial Alcohol . ...  :.......... ... .8vo,   4 00
* Kemp and Waugh's Landscape Gardening  (New Edition, Rewritten ) 12mo,   1 50
* McKay and Larsen's Principles and Practice of Butter-making ..... 8vo,   1 50
Maynard's Landscape Gardening as Applied to Home Decoration  . .12mo,   1 50
Record's Identification of the Economic Woods of the United States.  (In Press.)
Sanderson's Insects Injurious to Staple Crops ... ...   . ... .. 12mo,   1 50
    * Insect Pests of Farm, Garden, and Orchard.   . .   .Large 12mo,   3 00
* Schwarz's Longleaf Pine in Virgin Forest. ...... ... .,. .....·..12mo,   1 25
* Solotaroff's Field Book for Street-tree Mapping .. .... ....... 12mo,   0 75
    In lots of one dozen. ...   . . . . . . . .   .   8 00
    * Shade Trees in Towns and Cities... ...... ........... 8vo,   3 00
Stockbridge's Rocks and Soils .....  ........................8vo,   2 50
Winton's Microscopy of Vegetable Foods..................... ........ 8vo,   7 50
Woll's Handbook for Farmers and Dairymen ......................16mo,   1 50

## ARCHITECTURE.

* Atkinson's Orientation of Buildings or Planning for Sunlight.    8vo,   2 00
Baldwin's Steam Heating for Buildings.... .. . ... ......... ...12mo,   2 50
Berg's Buildings and Structures of American Railroads...............4to,   5 00

1

Birkmire's Architectural Iron and Steel............................8vo, $3 50
    Compound Riveted Girders as Applied in Buildings................8vo, 2 00
    Planning and Construction of High Office Buildings.... ........8vo, 3 50
    Skeleton Construction in Buildings... ..........................8vo, 3 00
Briggs's Modern American School Buildings............. ... . .8vo, 4 00
Byrne's Inspection of Materials and Workmanship Employed in Construction.
    16mo, 3 00
Carpenter's Heating and Ventilating of Buildings . ..................8vo, 4 00
* Corthell's Allowable Pressure on Deep Foundations...... ..........12mo, 1 25
* Eckel's Building Stones and Clays.. ... .. .. ... . 8vo, 3 00
Freitag's Architectural Engineering ...... .. ...................8vo, 3 50
    Fire Prevention and Fire Protection. (In Press.)
    Fireproofing of Steel Buildings .. ..... .. ..................8vo, 2 50
Gerhard's Guide to Sanitary Inspections. (Fourth Edition, Entirely Re-
    vised and Enlarged )..... ........... ...............12mo, 1 50
   * Modern Baths and Bath Houses .......... ... . ..........8vo, 3 00
    Sanitation of Public Buildings ............ ....... . ..... . 12mo, 1 50
    Theatre Fires and Panics ...... . .... .. 12mo, 1 50
   * The Water Supply, Sewerage and Plumbing of Modern City Buildings,
    8vo, 4 00
Johnson's Statics by Algebraic and Graphic Methods.... ... ... . 8vo, 2 00
Kellaway's How to Lay Out Suburban Home Grounds . .... 8vo, 2 00
Kidder's Architects' and Builders' Pocket-book...... 16mo, mor , 5 00
Merrill's Stones for Building and Decoration... . ...... . .8vo, 5 00
Monckton's Stair-building ... ... . .... ...... . ... 4to, 4 00
Patton's Practical Treatise on Foundations ...... . .. .. . 8vo, 5 00
Peabody's Naval Architecture . ........ ......... .. . .... 8vo, 7 50
Rice's Concrete-block Manufacture . ...................8vo, 2 00
Richey's Handbook for Superintendents of Construction . . 16mo, mor 4 00
    Building Foreman's Pocket Book and Ready Reference . 16mo, mor. 5 00
   * Building Mechanics' Ready Reference Series
    * Carpenters' and Woodworkers' Edition ........ .16mo, mor. 1 50
    * Cement Workers' and Plasterers' Edition . 16mo, mor 1 50
    * Plumbers', Steam-Fitters', and Tinners' Edition.. 16mo, mor 1 50
    * Stone- and Brick-masons' Edition . . ... ...16mo, mor 1 50
Sabin's House Painting . .. . . . . . ........... 12mo, 1 00
Siebert and Biggin's Modern Stone-cutting and Masonry...; .... . 8vo, 1 50
Snow's Principal Species of Wood ..... .. ... ...........8vo, 3 50
Wait's Engineering and Architectural Jurisprudence .. ........... 8vo, 6 00
    Sheep, 6 50
    Law of Contracts ...................... .. ..... .... ..... ..8vo, 3 00
    Law of Operations Preliminary to Construction in Engineering and
    Architecture ..................... . .. ....... 8vo, 5 00
    Sheep, 5 50
Wilson's Air Conditioning ....... ......... . .......... ... ...12mo, 1 50
Worcester and Atkinson's Small Hospitals, Establishment and Maintenance,
    Suggestions for Hospital Architecture, with Plans for a Small
    Hospital.......... ..... . ......... . ... ....12mo, 1 25


# ARMY AND NAVY.

Bernadou's Smokeless Powder, Nitro-cellulose, and the Theory of the Cellu-
    lose Molecule ...... .............. ... ........ ... .... . 12mo, 2 50
Chase's Art of Pattern Making ...... .. ............ . 12mo, 2 50
    Screw Propellers and Marine Propulsion.. . ........ .. .... . 8vo, 3 00
* Cloke's Enlisted Specialists' Examiner...... ......... . .... . 8vo, 2 00
    * Gunner's Examiner......... ........ ..... ...... . ..8vo, 1 50
Craig's Azimuth ....................... ... .... .:..........4to, 3 50
Crehore and Squier's Polarizing Photo-chronograph.. . .. ..........8vo, 3 00
* Davis's Elements of Law ....... ... :. ..................8vo, 2 50
    * Treatise on the Military Law of United States .... .... . .. 8vo, 7 00
* Dudley's Military Law and the Procedure of Courts-martial.. Large 12mo, 2 50
Durand's Resistance and Propulsion of Ships..... . .............. . 8vo, 5 00
* Dyer's Handbook of Light Artillery ...........................12mo, 3 00

Eissler's Modern High Explosives ....................... 8vo $4 00
* Fiebeger's Text-book on Field Fortification...... ..........Large 12mo, 2 00
Hamilton and Bond's The Gunner's Catechism............. ...........18mo, 1 00
* Hoff's Elementary Naval Tactics ..................... 8vo, 1 50
Ingalls's Handbook of Problems in Direct Fire ....................8vo, 4 00
    * Interior Ballistics. ....... .. . 8vo, 3 00
* Lissak's Ordnance and Gunnery , .... . . . .. ... 8vo, 6 00
* Ludlow's Logarithmic and Trigonometric Tables.. ... .8vo, 1 00
* Lyons's Treatise on Electromagnetic Phenomena Vols I and II 8vo,each, 6 00
* Mahan's Permanent Fortifications. (Mercur) ...... .. . 8vo. half mor. 7 50
Manual for Courts-martial . ... .. 16mo, mor. 1 50
* Mercur's Attack of Fortified Places...... ... . .. ... .... ...12mo, 2 00
    * Elements of the Art of War .... . ...... . 8vo, 4 00
Nixon's Adjutants' Manual . ..... ..... .. ...... .. 24mo, 1 00
Peabody's Naval Architecture .. .........,.... .. . . 8vo, 7 50
* Phelps's Practical Marine Surveying .... . .................8vo, 2 50
Putnam's Nautical Charts . . . . . ..........8vo, 2 00
Rust's Ex-meridian Altitude, Azimuth and Star-Finding Tables .... 8vo 5 00
* Selkirk's Catechism of Manual of Guard Duty . 24mo, 0 50
Sharpe's Art of Subsisting Armies in War . . ..18mo, mor. 1 50
* Taylor's Speed and Power of Ships 2 vols. Text 8vo, plates oblong 4to, 7 50
* Tupes and Poole's Manual of Bayonet Exercises and Musketry Fencing
    24mo, leather, 0 50
* Weaver's Military Explosives . . . ...... .. 8vo, 3 00
* Woodhull's Military Hygiene for Officers of the Line ..Large 12mo, 1 50

## ASSAYING.

Betts's Lead Refining by Electrolysis .. . .. ... . 8vo, 4 00
*Butler's Handbook of Blowpipe Analysis , ... ... .16mo, 0 75
Fletcher's Practical Instructions in Quantitative Assaying with the Blowpipe
    16mo, mor. 1 50
Furman and Pardoe's Manual of Practical Assaying . . 8vo, 3 00
Lodge's Notes on Assaying and Metallurgical Laboratory Experiments 8vo, 3 00
Low's Technical Methods of Ore Analysis . ......... .8vo, 3 00
Miller's Cyanide Process . . . .. .... . ...12mo, 1 00
    Manual of Assaying ... . 12mo, 1 00
Minet's Production of Aluminum and its Industrial Use (Waldo) . 12mo, 2 50
Ricketts and Miller's Notes on Assaying ... . .8vc 3 00
Robine and Lenglen's Cyanide Industry (Le Clerc) . . ... .8vo 4 00
* Seamon's Manual for Assayers and Chemists . Large 12mo, 2 50
Ulke's Modern Electrolytic Copper Refining . ... .. .. 8vo, 3 00
Wilson's Chlorination Process .. .. ... . .. .. 12mo, 1 50
    Cyanide Processes ............... .. ... ... .12mo, 1 50

## ASTRONOMY.

Comstock's Field Astronomy for Engineers..... .. .... ...... .8vo, 2 50
Craig's Azimuth.... . ..... .. ........ . .4to, 3 50
Crandall's Text-book on Geodesy and Least Squares .......... . 8vo, 3 00
Doolittle's Treatise on Practical Astronomy . .. ...... .. ... 8vo, 4 00
Hayford's Text-book of Geodetic Astronomy .... ...... . .. ....8vo, 3 00
Hosmer's Azimuth .. . . . ... ....16mo, mor 1 00
    * Text-book on Practical Astronomy . .... 8vo, 2 00
Merriman's Elements of Precise Surveying and Geodesy. .... ...... 8vo, 2 50
* Michie and Harlow's Practical Astronomy . . ......8vo, 3 00
Rust's Ex-meridian Altitude, Azimuth and Star-Finding Tables ..... 8vo, 5 00
* White's Elements of Theoretical and Descriptive Astronomy........12mo, 2 00

## CHEMISTRY.

* Abderhalden's Physiological Chemistry in Thirty Lectures. (Hall and
    Defren.). ... ........ .... .. .8vo, 5 00
* Abegg's Theory of Electrolytic Dissociation (von Ende) . .....12mo, 1 25
Alexeyeff's General Principles of Organic Syntheses (Matthews.)..... 8vo, 3 00

4

5

## CIVIL ENGINEERING.

### BRIDGES AND ROOFS.   HYDRAULICS   MATERIALS OF ENGINEER-ING.   RAILWAY ENGINEERING.

Ogden and Cleveland's Practical Methods of Sewage Disposal for Residences, Hotels, and Institutions. (In Press.)
Parsons's Disposal of Municipal Refuse .... .......................8vo, $2 00
Patton's Treatise on Civil Engineering . . . ........ 8vo, half leather, 7 50
Reed's Topographical Drawing and Sketching . ...... ............4to, 5 00
Riemer's Shaft-sinking under Difficult Conditions (Corning and Peele.).8vo, 3 00
Siebert and Biggin's Modern Stone-cutting and Masonry ........ ....8vo, 1 50
Smith's Manual of Topographical Drawing (McMillan.).... .. .....8vo, 2 50
Soper's Air and Ventilation of Subways .........................12mo, 2 50
* Tracy's Exercises in Surveying............................12mo, mor. 1 00
Tracy's Plane Surveying . ...... ........................16mo, mor. 3 00
Venable's Garbage Crematories in America............. ...........8vo, 2 00
    Methods and Devices for Bacterial Treatment of Sewage............8vo, 3 00
Wait's Engineering and Architectural Jurisprudence...................8vo, 6 00
                                                              Sheep, 6 50
    Law of Contracts..... .......................................8vo, 3 00
    Law of Operations Preliminary to Construction in Engineering and
        Architecture ..... .... .........................................8vo, 5 00
                                                              Sheep, 5 50
Warren's Stereotomy—Problems in Stone-cutting .....................8vo, 2 50
* Waterbury's Vest-Pocket Hand-book of Mathematics for Engineers
                                          $2\frac{7}{8} \times 5\frac{3}{8}$ inches, mor. 1 00
        * Enlarged Edition, Including Tables... .......... ........ ...mor 1 50
Webb's Problems in the Use and Adjustment of Engineering Instruments.
                                                         16mo, mor. 1 25
Wilson's Topographic Surveying ............................. .8vo, 3 50

## BRIDGES AND ROOFS.

Boller's Practical Treatise on the Construction of Iron Highway Bridges 8vo, 2 00
        * Thames River Bridge ....... .. .. .......... .Oblong paper, 5 00
Burr and Falk's Design and Construction of Metallic Bridges....... .. 8vo, 5 00
        'Influence Lines for Bridge and Roof Computations ... ........ ..... .8vo, 3 00
Du Bois's Mechanics of Engineering Vol II.. ... ............Small 4to, 10 00
Foster's Treatise on Wooden Trestle Bridges ..................... . 4to, 5 00
Fowler's Ordinary Foundations... .. ... 8vo, 3 50
Greene's Arches in Wood, Iron, and Stone ...... .............8vo, 2 50
        Bridge Trusses.. ... .. ..... ........................8vo, 2 50
        Roof Trusses . .. .... .. .....................8vo, 1 25
Grimm's Secondary Stresses in Bridge Trusses ...... .......... .8vo, 2 50
Heller's Stresses in Structures and the Accompanying Deformations ...8vo, 3 00
Howe's Design of Simple Roof-trusses in Wood and Steel.... ...... .8vo 2 00
        Symmetrical Masonry Arches .... .....................8vo, 2 50
        Treatise on Arches .. . ..... .. .... .. ...... 8vo, 4 00
* Hudson's Deflections and Statically Indeterminate Stresses. . .Small 4to, 3 50
        * Plate Girder Design ... . . . . . . . . . . . . 8vo, 1 50
* Jacoby's Structural Details, or Elements of Design in Heavy Framing, 8vo, 2 25
Johnson, Bryan and Turneaure's Theory and Practice in the Designing of
        Modern Framed Structures ......... ............ ....Small 4to, 10 00
*¯Johnson, Bryan and Turneaure's Theory and Practice in the Designing of
        Modern Framed Structures New Edition. Part I ........8vo, 3 00
        * Part II. New Edition . ... ...... .......... . ...... ....8vo, 4 00
Merriman and Jacoby's Text-book on Roofs and Bridges
        Part I. Stresses in Simple Trusses.. .... ...... ... ......8vo, 2 50
        Part II. Graphic Statics . . ...... ...... .. . ...8vo, 2 50
        Part III Bridge Design ... .. ...... . . .. . . .8vo, 2 50
        Part IV Higher Structures ... ..... .. ..........8vo, 2 50
Ricker's Design and Construction of Roofs (In Press.)
Sondericker's Graphic Statics, with Applications to Trusses, Beams, and
        Arches .. ... .. ..... . .................... .... ..8vo, 2 00
Waddell's De Pontibus, Pocket-book for Bridge Engineers ... . 16mo, mor. 2 00
        * Specifications for Steel Bridges .. ... ...... . ..... .12mo, . 50
Waddell and Harrington's Bridge Engineering (In Preparation.)

## HYDRAULICS.

Barnes's Ice Formation ............................. ...................8vo, 3 00
Bazin's Experiments upon the Contraction of the Liquid Vein Issuing from
        an Orifice. (Trautwine )....... ....... .. .............8vo, 2 00

7

Bovey's Treatise on Hydraulics ... ..............................................8vo, $5 00
Church's Diagrams of Mean Velocity of Water in Open Channels
                                      Oblong 4to, paper, 1 50
    Hydraulic Motors . ....................................... . ... . 8vo, 2 00
    Mechanics of Fluids (Being Part IV of Mechanics of Engineering) 8vo, 3 00
Coffin's Graphical Solution of Hydraulic Problems. .........16mo, mor 2 50
Flather's Dynamometers, and the Measurement of Power......... . 12mo, 3 00
Folwell's Water-supply Engineering. ...................... ....... 8vo, 4 00
Frizell's Water-power ... . ................................ .. 8vo, 5 00
Fuertes's Water and Public Health . .......................12mo, 1 50
    Water-filtration Works . ... .................. ...........12mo, 2 50
Ganguillet and Kutter's General Formula for the Uniform Flow of Water in
    Rivers and Other Channels (Hering and Trautwine ) .. .8vo, 4 00
Hazen's Clean Water and How to Get It ...................Large 12mo, 1 50
    Filtration of Public Water-supplies . ......... ..... . . 8vo, 3 00
Hazelhurst's Towers and Tanks for Water-works..... ....... ...8vo, 2 50
Herschel's 115 Experiments on the Carrying Capacity of Large, Riveted, Metal
    Conduits .. . ........ ......,................. . .8vo, 2 00
Hoyt and Grover's River Discharge . ..................... .8vo, 2 00
Hubbard and Kiersted's Water-works Management and Maintenance
                                              8vo, 4 00
* Lyndon's Development and Electrical Distribution of Water Power
                                              8vo, 3 00
Mason's Water-supply (Considered Principally from a Sanitary Stand-
    point ). ....... ..... . ........ ...... ... .... . 8vo, 4 00
* Merriman's Treatise on Hydraulics. 9th Edition, Rewritten . 8vo, 4 00
* Molitor's Hydraulics of Rivers, Weirs and Sluices ...... ... .... 8vo, 2 00
* Morrison and Brodie's High Masonry Dam Design .. . . . . . .8vo, 1 50
* Richards's Laboratory Notes on Industrial Water Analysis .. .. 8vo, 50
Schuyler's Reservoirs for Irrigation, Water-power, and Domestic Water-
    supply Second Edition, Revised and Enlarged...... Large 8vo, 6 00
* Thomas and Watt's Improvement of Rivers ........ ........ .....4to, 6 00
Turneaure and Russell's Public Water-supplies ... ... ...... 8vo, 5 00
* Wegmann's Design and Construction of Dams 6th Ed , enlarged . 4to, 6 00
    Water-Supply of the City of New York from 1658 to 1895 .... 4to, 10 00
Whipple's Value of Pure Water .... .. .. ...............Large 12mo, 1 00
Williams and Hazen's Hydraulic Tables . ........................8vo, 1 50
Wilson's Irrigation Engineering ..... . ................8vo, 4 00
Wood's Turbines ....... . . ...... ........................8vo, 2 50

## MATERIALS OF ENGINEERING.

Baker's Roads and Pavements .....................................8vo, 5 00
    Treatise on Masonry Construction..............................8vo, 5 00
Black's United States Public Works .. .................... Oblong 4to, 5 00
* Blanchard and Drowne's Highway Engineering, as Presented at the
    Second International Road Congress, Brussels, 1910 ........8vo, 2 00
Bleininger's Manufacture of Hydraulic Cement (In Preparation )
* Bottler's German and American Varnish Making. (Sabin.). Large 12mo. 3 50
Burr's Elasticity and Resistance of the Materials of Engineering . . . ..8vo, 7 50
Byrne's Highway Construction.. ..................................8vo, 5 00
    Inspection of the Materials and Workmanship Employed in Construction.
                                      16mo, 3 00
Church's Mechanics of Engineering...................................8vo, 6 00
    Mechanics of Solids (Being Parts I, II, III of Mechanics of Engineer-
    ing . . . ' . . ........ . ........ ......8vo, 4 50
Du Bois's Mechanics of Engineering
    Vol I Kinematics, Statics, Kinetics.... ............. .Small 4to, 7 50
    Vol II The Stresses in Framed Structures, Strength of Materials and
    Theory of Flexures .. .. .. ......./..... .....Small 4to, 10 00
* Eckel's Building Stones and Clays . . . ...... .. 8vo, 3 00
    * Cements, Limes, and Plasters.. ... ..... .............. ...8vo, 6 00
Fowler's Ordinary Foundations . ..................................8vo, 3 50
* Greene's Structural Mechanics ........ ... ... ...... ...... .........8vo, 2 50
Holley's Analysis of Paint and Varnish Products. ' (In Press.)
    * Lead and Zinc Pigments......................?.....Large 12mo, 3 00

8

## RAILWAY ENGINEERING.

## DRAWING

Wilson's (H. M ) Topographic Surveying. .... ...   ... ..... ...  .8vo,  $3 50
* Wilson's (V. T ) Descriptive Geometry. .. ........ ... ... ... ...  8vo,  1 50
    Free-hand Lettering  .... ..... ..•••• ..   .   ... ...8vo,  1 00
    Free-hand Perspective   ....  •••••• ...  ... ...  ...... 8vo,  2 50
Woolf's Elementary Course in Descriptive Geometry.  .. ... .Large 8vo,  3 00

# ELECTRICITY AND PHYSICS.

* Abegg's Theory of Electrolytic Dissociation   (von Ende )... .. 12mo,  1 25
Andrews's Hand-book for Street Railway Engineers    .3×5 inches mor.  1 25
Anthony and Ball's Lecture-notes on the Theory of Electrical Measure-
    ments ........ .•......... ... .... .. .   .  12mo,  1 00
Anthony and Brackett's Text-book of Physics   (Magie )   Large 12mo,  3 00
Benjamin's History of Electricity ... ........ ..   ......... ...  .8vo,  3 00
Betts's Lead Refining and Electrolysis .'  . ...  . ... ... 8vo,  4 00
* Burgess and Le Chatelier's Measurement of High Temperatures   Third
    Edition....   .   8vo,  4 00
Classen's Quantitative Chemical Analysis by Electrolysis   (Boltwood ) 8vo,  3 00
* Collins's Manual of Wireless Telegraphy and Telephony  .. .... 12mo,'  1 50
Crehore and Squier's Polarizing Photo-chronograph . ...   . ...... 8vo,  3 00
* Danneel's Electrochemistry.  (Merriam ) ..   .   12mo,  1 25
Dawson's "Engineering" and Electric Traction Pocket-book  . 16mo, mor  5 00
Dolezalek's Theory of the Lead Accumulator (Storage Battery)   (von Ende )
    12mo,  2 50
Duhem's Thermodynamics and Chemistry   (Burgess )  ... ....   8vo,  4 00
Flather's Dynamometers, and the Measurement of Power ..  .. 12mo,  3 00
* Getman's Introduction to Physical Science.. . .. ..   ..  .12mo,  1 50
Gilbert's De Magnete   (Mottelay )...   ...   .   .. .. 8vo,  2 50
* Hanchett's Alternating Currents   .   . ... 12mo,  1 00
Hering's Ready Reference Tables (Conversion Factors)  .. .16mo, mor  2 50
* Hobart and Ellis's High-speed Dynamo Electric Machinery   ..  8vo,  6 00
Holman's Precision of Measurements   ...... .  ..   .  8vo,  2 00
    Telescope-Mirror-scale Method, Adjustments, and Tests   Large 8vo,  0 75
* Hutchinson's High-Efficiency Electrical Illuminants and Illumination.
    .   .   Large 12mo,  2 50
* Jones's Electric Ignition  .. . .   .   8vo,  4 00
Karapetoff's Experimental Electrical Engineering
    * Vol  I   ......   .   . ...  .   8vo,  3 50
    * Vol. II   . .   .. . 8vo,  2 50
Kinzbrunner's Testing of Continuous-current Machines .   .   8vo,  2 00
Landauer's Spectrum Analysis.  (Tingle )  .... .   . 8vo,  3 00
Lob's Electrochemistry of Organic Compounds   (Lorenz )  .   8vo,  3 00
* Lyndon's Development and Electrical Distribution of Water Power  8vo,  3 00
* Lyons's Treatise on Electromagnetic Phenomena  Vols, I and II 8vo, each,  6 00
* Michie's Elements of Wave Motion Relating to Sound and Light .  8vo,  4 00
* Morgan's Physical Chemistry for Electrical Engineers  . . ... 12mo,  1 50
* Norris's Introduction to the Study of Electrical Engineering  .. 8vo,  2 50
Norris and Dennison's Course of Problems on the Electrical Characteristics of
    Circuits and Machines   (In Press )
* Parshall and Hobart's Electric Machine Design. . ....  4to, half mor,  12 50
Reagan's Locomotives. Simple, Compound, and Electric   New Edition
    Large 12mo,  3 50
* Rosenberg's Electrical Engineering   (Haldane Gee—Kinzbrunner )..8vo,  2 00
* Ryan's Design of Electrical Machinery:
    * Vol. I. Direct Current Dynamos   .. . . .  ...... 8vo,  1 50
    Vol. II. Alternating Current Transformers.  (In Press )
    Vol. III. Alternators, Synchronous Motors, and Rotary Convertors.
    (In Preparation )
Ryan, Norris, and Hoxie's Text Book of Electrical Machinery.. .  .. 8vo,  2 50
Schapper's Laboratory Guide for Students in Physical Chemistry   12mo,  1 00
* Tillman's Elementary Lessons in Heat.................. .'..... ...8vo,  1 50
* Timbie's Elements of Electricity.............   ........Large 12mo,  2 00
    * Answers to Problems in Elements of Electricity........12mo, Paper  0 25
Tory and Pitcher's Manual of Laboratory Physics ...........Large 12mo,  2 00
Ulke's' Modern Electrolytic Copper Refining .......................8vo,  3 00
* Waters's Commercial Dynamo Design .  .... ..... ........ ....8vo,  2 00

# LAW.

* Brennan's Hand-book of Useful Legal Information for Business Men

            16mo, mor. $5 00
* Davis's Elements of Law .   . . . . . . . .   . . . . . . . . . . . .8vo,   2 50
  * Treatise on the Military Law of United States   . . .   8vo,   7 00
* Dudley's Military Law and the Procedure of Courts-martial   Large 12mo,   2 50
Manual for Courts-martial   ..    . . . . . .   .   . . . . .16mo, mor   1 50
Wait's Engineering and Architectural Jurisprudence . .   . .   . . . . . . .8vo,   6 00
                    Sheep,   6 50
  Law of Contracts . . .   . . . . . .   . . . . . . . . . .   . . . . 8vo,   3 00
  Law of Operations Preliminary to Construction in Engineering and
   Architecture .   . . . . . . . .   . . . . . . .   . . . . . . .   . . . . . . . . . . 8vo,   5 00
                    Sheep,   5 50

# MATHEMATICS.

Baker's Elliptic Functions .   ..   . . . .   . . . . . .   .   .   . 8vo,   1 50
Briggs's Elements of Plane Analytic Geometry   (Bôcher )   ..12mo,   1 00
* Buchanan's Plane and Spherical Trigonometry   ..   .   ..   . .8vo,   1 00
Byerly's Harmonic Functions   . . . . . . . .   . . .   . . . . . 8vo,   1 00
Chandler's Elements of the Infinitesimal Calculus . . .   . . .   . .12mo,   2 00
* Coffin's Vector Analysis   ..   .   . . . . . . . . . .   .   . 12mo,   2 50
Compton's Manual of Logarithmic Computations . . . . . . . . . . .   12mo,   1 50
* Dickson's College Algebra .   . . . . .   .    . . . . . . . . . . . .Large 12mo,   1 50
  * Introduction to the Theory of Algebraic Equations   . .Large 12mo,   1 25
Emch's Introduction to Projective Geometry and its Application . . . .   8vo,   2 50
Fiske's Functions of a Complex Variable . . . .   . . .   . . .   . . .8vo,   1 00
Halsted's Elementary Synthetic Geometry   . . . .   .   . 8vo,   1 50
  Elements of Geometry. . . . . . . . . . . . .   . .   . . . . . . .   . 8vo,   1 75
  * Rational Geometry   . . . . .   .    .   . . . .   12mo,   1 50
  Synthetic Projective Geometry    .   .. 8vo,   1 00
* Hancock's Lectures on the Theory of Elliptic Functions   8vo,   5 00
Hyde's Grassmann's Space Analysis   .. . . .   ..   . . . .8vo,   1 00
* Johnson's (J B.) Three-place Logarithmic Tables   Vest-pocket size, paper,   0 15
                  * 100 copies,   5 00
      * Mounted on heavy cardboard, 8 × 10 inches,   0 25
                  * 10 copies,   2 00
Johnson's (W W ) Abridged Editions of Differential and Integral Calculus
                Large 12mo, 1 vol   2 50
  Curve Tracing in Cartesian Co-ordinates ..   . . . . . .   .   . 12mo,   1 00
  Differential Equations ..   .   . . .   ..   8vo,   1 00
  Elementary Treatise on Differential Calculus   .   Large 12mo,   1 50
  Elementary Treatise on the Integral Calculus   .   . . .   Large 12mo,   1 50
  * Theoretical Mechanics . . .   . . . . . . . . . .   .   . . . . . . 12mo,   3 00
  Theory of Errors and the Method of Least Squares. . .   .   . . .12mo,   1 50
  Treatise on Differential Calculus. . . . . . . . .   . . . . . . . . . . . Large 12mo,   3 00
  Treatise on the Integral Calculus   . . . . . . . . .   . . . . . . . Large 12mo,   3 00
  Treatise on Ordinary and Partial Differential Equations   .   Large 12mo,   3 50
Karapetoff's Engineering Applications of Higher Mathematics.
  * Part I. Problems on Machine Design. . . . . . .   Large 12mo,   0 75
Koch's Practical Mathematics   (In Press.)
Laplace's Philosophical Essay on Probabilities   (Truscott and Emory ) . 12mo,   2 00
* Le Messurier's Key to Professor W. W. Johnson's Differential Equations.
                   Small 8vo,   1 75
* Ludlow's Logarithmic and Trigonometric Tables . . . . . . .   . . . . . . . .8vo,   1 00
* Ludlow and Bass's Elements of Trigonometry and Logarithmic and Other
  Tables . . . . . . . . .   .   . . . . . . . .   . . . . . . .   .   ..   . . . . 8vo,   3 00
  * Trigonometry and Tables published separately   . . . . .Each,   2 00
Macfarlane's Vector Analysis and Quaternions . . .   .   .. . . . . . . .8vo,   1 00
McMahon's Hyperbolic Functions   . . . . . .   . . . . .8vo,   1 00
Manning's Irrational Numbers and their Representation by Sequences and
  Series. . . .   . . . . .   ..   . .   .   .   . . .12mo,   1 25
* Martin's Text Book on Mechanics.   Vol. I. Statics    .12mo,   1 25
  * Vol. II.   Kinematics and Kinetics   .   .   12mo,   1 50
  * Vol. III.   Mechanics of Materials..   ..   .   .   . 12mo,   1 50

Mathematical Monographs   Edited by Mansfield Merriman and Robert
    S Woodward   .   ... ......   .   Octavo, each $1 00
    No. 1  History of Modern Mathematics, by David Eugene Smith
    No 2  Synthetic Projective Geometry, by George Bruce Halsted
    No 3  Determinants, by Laenas Gifford Weld   No 4. Hyper-
    bolic Functions, by James McMahon   No. 5  Harmonic Func-
    tions, by William E. Byerly.   No 6 Grassmann's Space Analysis,
    by Edward W Hyde   No. 7  Probability and Theory of Errors,
    by Robert S Woodward. No. 8   Vector Analysis and Quaternions,
    by Alexander Macfarlane   No. 9  Differential Equations, by
    William Woolsey Johnson   No 10  The Solution of Equations,
    by Mansfield Merriman.   No 11 Functions of a Complex Variable,
    by Thomas S Fiske
Maurer's Technical Mechanics   . .   .   .. 8vo,   4 00
Merriman's Method of Least Squares   . .   . .8vo,   2 00
    Solution of Equations .. .   .   ... 8vo,   1 00
* Moritz's Elements of Plane Trigonometry.   .   8vo,   2 00
Rice and Johnson's Differential and Integral Calculus   2 vols in one
    Large 12mo,   1 50
    Elementary Treatise on the Differential Calculus   .   Large 12mo,   3 00
Smith's History of Modern Mathematics   .. .   8vo,   1 00
* Veblen and Lennes's Introduction to the Real Infinitesimal Analysis of One
    Variable .. ......   .. .. .. .. ....   .... ...8vo,   2 00
* Waterbury's Vest Pocket Hand-book of Mathematics for Engineers
    2⅝ × 5⅜ inches, mor   1, 00
    * Enlarged Edition, Including Tables .. . . . . . . .   . mor   1 50
Weld's Determinants .. . .   .. . .. .... . . ...   8vo,   1 00
Wood's Elements of Co-ordinate Geometry ... . . . . . . . .   . ...   8vo,   2 00
Woodward's Probability and Theory of Errors . . . . . . . . . .   . . . . . . ...   8vo,   1 00

## MECHANICAL ENGINEERING.
MATERIALS OF ENGINEERING. STEAM-ENGINES AND BOILERS.

Bacon's Forge Practice . . . . .   . . .   . . . . . .   . . . .   .   12mo,   1 50
Baldwin's Steam Heating for Buildings .. . ..   .   .   12mo,   2 50
Barr and Wood's Kinematics of Machinery   . . . . . . . . ... .   . 8vo,   2 50
* Bartlett's Mechanical Drawing   . . . . . . . . . . ..   . . . . . . .. . .   8vo,   3 00
*   "   "   "   Abridged Ed .. .   . .   . .   ...   8vo,   1 50
* Bartlett and Johnson's Engineering Descriptive Geometry   8vo,   1 50
* Burr's Ancient and Modern Engineering and the Isthmian Canal   .   8vo,   3 50
Carpenter's Heating and Ventilating Buildings   . . .   8vo,   4 00
* Carpenter and Diederichs's Experimental Engineering   .   8vo,   6 00
* Clerk's The Gas, Petrol and Oil Engine . . . . . . . . . . .   .   8vo,   4 00
Compton's First Lessons in Metal Working . . . . . . . . .   .   12mo,   1 50
Compton and De Groodt's Speed Lathe .. . ...   .   . . . . 12mo,   1 50
Coolidge's Manual of Drawing   . . . .   ..   8vo, paper,   1 00
Coolidge and Freeman's Elements of General Drafting for Mechanical En-
    gineers   . . . . . . . . .   . .   .   Oblong 4to,   2 50
Cromwell's Treatise on Belts and Pulleys ..   12mo,   1 50
    Treatise on Toothed Gearing   .   . .   12mo,   1 50
Dingey's Machinery Pattern Making ..   . . .   .   12mo,   2 00
Durley's Kinematics of Machines. . . . .   .   8vo,   4 00
Flanders's Gear-cutting Machinery .   .   .   Large 12mo,   3 00
Flather's Dynamometers and the Measurement of Power   . . 12mo,   3 00
    Rope Driving . . . . . .. . . . . . .   . . . .   .   J2mo,   2 00
Gill's Gas and Fuel Analysis for Engineers . . .   ...   . ..12mo,   1 25
Goss's Locomotive Sparks . . . . . . . . . .   . . . .   .. .. . . .   .. 8vo,   2 00
* Greene's Pumping Machinery . . . . .   . . . . . . . .   .   8vo,   4 00
Hering's Ready Reference Tables (Conversion Factors). . . .   .   16mo, mor   2 50
‡ Hobart and Ellis's High Speed Dynamo Electric Machinery   8vo,   6 00
Hutton's Gas Engine . . . . . . . . . . . . . . . . . .   ...   .. 8vo,   5 00
Jamison's Advanced Mechanical Drawing . ..   .   . .8vo,   2 00
    Elements of Mechanical Drawing.. .   . .   .   8vo,   2 50
Jones's Gas Engine . . . . . . . . . .   .   ...   ...   ..8vo,   4 00
    Machine Design:
    Part I.   Kinematics of Machinery   8vo,   1 50
    Part II.   Form, Strength, and Proportions of Parts. . . . . .   ..8vo,   3 00

13

```
* Kaup's Machine Shop Practice      .     .     .          Large 12mo  $1 25
* Kent's Mechanical Engineer's Pocket-Book      .      ... 16mo, mor.   5 00
Kerr's Power and Power Transmission ..     .        .... .... .  8vo,   2 00
* Kimball and Barr's Machine Design. .      .      .      .   . 8vo,    3 00
* King's Elements of the Mechanics of Materials and of Power of Trans-
        mission ...            ..    .             . ..8vo,             2 50
* Lanza's Dynamics of Machinery.    .     .      .     .     . 8vo,     2 50
Leonard's Machine Shop Tools and Methods.   .   .  .   .......8vo,      4 00
* Levin's Gas Engine ....  ..  ..   ....  ..      .    .   . ... 8vo,   4 00
* Lorenz's Modern Refrigerating Machinery.  (Pope, Haven, and Dean) 8vo,  4 00
MacCord's Kinematics, or, Practical Mechanism     .   .   .   . 8vo,    5 00
        Mechanical Drawing  ....  .   ..  ......  .   ..  ... 4to,      4 00
        Velocity Diagrams  .     .     .    .   .    . ... 8vo,         1 50
MacFarland's Standard Reduction Factors for Gases ...  ..  .. 8vo,      1 50
Mahan's Industrial Drawing   (Thompson)..... ..... .. ...' .. 8vo,      3 50
Mehrtens's Gas Engine Theory and Design ..  ..  ....  Large 12mo,      2 50
Miller, Berry, and Riley's Problems in Thermodynamics and Heat Engineer-
        ing..  .     .    ...  ......  ..  ......  ".  8vo, paper,      0 75
Oberg's Handbook of Small Tools. ..  ....  .......    Large 12mo,       2 50
* Parshall and Hobart's Electric Machine Design. Small 4to, half leather, 12 50
* Peele's Compressed Air Plant.  Second Edition, Revised and Enlarged 8vo,  3 50
* Perkins's Introduction to General Thermodynamics  .    .    12mo.     1 50
Poole's Calorific Power of Fuels ....  ......  .   ....  .... . 8vo,    3 00
* Porter's Engineering Reminiscences, 1855 to 1882 . ...  ....  . . 8vo,  3 00
Randall's Treatise on Heat.   (In Press.)
* Reid's Mechanical Drawing    (Elementary and Advanced )'   .   . 8vo,  2 00
        Text-book of Mechanical Drawing and Elementary Machine Design 8vo,  3 00
Richards's Compressed Air ...  ..  ....  ...  ... .  ... 12mo,          1 50
Robinson's Principles of Mechanism ....  .  ..  .   . ... 8vo,          3 00
Schwamb and Merrill's Elements of Mechanism...  ..  .   .    . 8vo,     3 00
Smith (A. W ) and Marx's Machine Design.  ....................8vo,      3 00
Smith's (O ) Press-working of Metals .   .    .   .. .  ...    ..8vo,   3 00
Sorel's Carbureting and Combustion in Alcohol Engines   (Woodward and
        Preston )  ......  ..  ....   .          . ..Large 12mo,       3 00
Stone's Practical Testing of Gas and Gas Meters        .    8vo,       3 50
Thurston's Animal as a Machine and Prime Motor, and the Laws of Energetics
                                                        12mo,          1 00
        Treatise on Friction and Lost Work in Machinery and Mill Work  8vo,  3 00
* Tillson's Complete Automobile Instructor  ...   .    .  . 16mo,       1 50
* Titsworth's Elements of Mechanical Drawing.   ...   .  Oblong 8vo,    1 25
Warren's Elements of Machine Construction and Drawing    .   .. 8vo,    7 50
* Waterbury's Vest Pocket Hand-book of Mathematics for Engineers.
                                        2⅞ × 5⅜ inches, mor            1 00
        * Enlarged Edition, Including Tables ...  ..  .            .mor.  1 50
Weisbach's Kinematics and the Power of Transmission  (Herrmann—
        Klein )   ...   .    ........  ....  .     .     . 8vo,        5 00
        Machinery of Transmission and Governors   (Hermann—Klein ) 8vo,  5 00
Wood's Turbines. ..  ..............  .  ....   ..  ......  .... 8vo,    2 50
```

## MATERIALS OF ENGINEERING.

```
Burr's Elasticity and Resistance of the Materials of Engineering. ...  ..8vo,  7 50
Church's Mechanics of Engineering ....  ..  .  .......  .... 8vo,       6 00
        Mechanics of Solids (Being Parts I, II, III of Mechanics of Engineering).
                                                        8vo,           4 50
* Greene's Structural Mechanics.   ..  .  ..  .  .     .  .......8vo,   2 50
Holley's Analysis of Paint and Varnish Products.   (In Press )
        * Lead and Zinc Pigments. .    .              ... Large 12mo,  3 00
Johnson's (C. M ) Rapid Methods for the Chemical Analysis of Special
        Steels, Steel-Making Alloys and Graphite ..  ... Large 12mo,   3 00
Johnson's (J B.) Materials of Construction. .   ..  ..  ..  ..  .8vo,   6 00
Keep's Cast Iron. ....  .     .     .     .    .    .8vo,               2 50
* King's Elements of the Mechanics of Materials and of Power of Trans-
        mission...  ........  ......  ..      ..  ..  ...  ..8vo,       2 50
Lanza's Applied Mechanics.   ....  ...    .     ......  ...  .  .8vo,   7 50
Lowe's Paints for Steel Structures  ......  ...  ...  ...  ......12mo,  1 00
Maire's Modern Pigments and their Vehicles  .  ...  ....  ...  ......12mo,  2 00
```

Maurer's Technical Mechanics . .. . ...........8vo, $4 00
Merriman's Mechanics of Materials ... . . . .. .. 8vo, 5 00
　　* Strength of Materials . . . . . .. . ....12mo, 1 00
Metcalf's Steel A Manual for Steel-users .. ... ... .. 12mo, 2 00
* Murdock's Strength of Materials 12mo, 2 00
Sabin's Industrial and Artistic Technology of Paint and Varnish . 8vo, 3 00
Smith's (A. W.) Materials of Machines .... ..... ... 12mo, 1 00
* Smith's (H E ) Strength of Material ... . .......... .. 12mo, 1 25
Thurston's Materials of Engineering... . . ... .. ...3 vols , 8vo, 8 00
　　Part I. Non-metallic Materials of Engineering, ... ... . . 8vo, 2 00
　　Part II Iron and Steel . . . . .. .. .. 8vo, 3 50
　　Part III A Treatise on Brasses, Bronzes, and Other Alloys and their
　　　Constituents .. . ... .... .. . . 8vo, 2 50
Waterbury's Laboratory Manual for Testing Materials of Construction.
　　　(In Press.)
Wood's (De V ) Elements of Analytical Mechanics .... .... 8vo, 3 00
　　Treatise on the Resistance of Materials and an Appendix on the
　　　Preservation of Timber . . .......... .. .. .. 8vo, 2 00
Wood's (M P ) Rustless Coatings· Corrosion and Electrolysis of Iron and
　　　Steel.. ................................. 8vo, 4 00

## STEAM-ENGINES AND BOILERS.

Berry's Temperature-entropy Diagram. Third Edition Revised and En-
　　larged ......... ............ .... 12mo, 2 50
Carnot's Reflections on the Motive Power of Heat. (Thurston ). . 12mo, 1 50
Chase's Art of Pattern Making ... ..... ..... ... .. ..... .12mo, 2 50
Creighton's Steam-engine and other Heat Motors .. .. ...... .8vo, 5 00
Dawson's "Engineering" and Electric Traction Pocket-book. ..16mo, mor 5 00
* Gebhardt's Steam Power Plant Engineering .......... .. 8vo, 6 00
Goss's Locomotive Performance . ..... ...... ..... 8vo, 5 00
Hemenway's Indicator Practice and Steam-engine Economy .. .12mo, 2 00
Hirshfeld and Barnard's Heat Power Engineering. (In Press.)
Hutton's Heat and Heat-engines . .. . 8vo, 5 00
　　Mechanical Engineering of Power Plants. .... . .... . . 8vo, 5 00
Kent's Steam Boiler Economy .. . .... ....... .. 8vo, 4 00
Kneass's Practice and Theory of the Injector ...... ........ . . 8vo, 1 50
MacCord's Slide-valves ........ .. ..... ..... ...8vo, 2 00
Meyer's Modern Locomotive Construction .. . ..... .. ...4to, 10 00
Miller, Berry, and Riley's Problems in Thermodynamics. .. 8vo, paper, 0 75
Moyer's Steam Turbine. . . . ... .................... 8vo, 4 00
Peabody's Manual of the Steam-engine Indicator... .. .. 12mo, 1 50
　　Tables of the Properties of Steam and Other Vapors and Temperature-
　　　Entropy Table. .... .. . ... . ... .. ...... .. ...8vo, 1 00
　　Thermodynamics of the Steam-engine and Other Heat-engines . . .8vo, 5 00
　　* Thermodynamics of the Steam Turbine ... . ...... .. . . 8vo, 3 00
　　Valve-gears for Steam-engines .... ...... ...... ...... 8vo, 2 50
Peabody and Miller's Steam-boilers ... ..8vo, 4 00
* Perkins's Introduction to General Thermodynamics 12mo. 1 50
Pupin's Thermodynamics of Reversible Cycles in Gases and Saturated Vapors
　　　(Osterberg ) .. ... .. ....... .. ..... 12mo, 1 25
Reagan's Locomotives. Simple, Compound, and Electric. New Edition.
　　　Large 12mo, 3 50
Sinclair's Locomotive Engine Running and Management .... .....12mo, 2 00
Smart's Handbook of Engineering Laboratory Practice . ........ .12mo, 2 50
Snow's Steam-boiler Practice . .................... ... ...8vo, 3 00
Spangler's Notes on Thermodynamics. .. . ........ .. ...... 12mo, 1 00
　　Valve-gears. ...... .. . ..... .. ..... ...8vo, 2 50
Spangler, Greene, and Marshall's Elements of Steam-engineering . . .8vo, 3 00
Thomas's Steam-turbines ... . .. .. ...... . ....... 8vo, 4 00
Thurston's Handbook of Engine and Boiler Trials, and the Use of the Indi-
　　cator and the Prony Brake.. ..... .. ...... . ...... 8vo, 5 00
　　Handy Tables ... . .. . .. .. .. ...... . 8vo, 1 50
　　Manual of Steam-boilers, their Designs, Construction, and Operation 8vo, 5 00
　　Manual of the Steam-engine .... .. . ............2 vols., 8vo, 10 00
　　　Part I History, Structure, and Theory .............. ....8vo, 6 00
　　　Part II Design, Construction, and Operation..............8vo, 6 00

**Wehrenfennig's** Analysis and Softening of Boiler Feed-water. (Patterson )

8vo, $4 00

**Weisbach's** Heat, Steam, and Steam-engines. (Du Bois )........ ...8vo, 5 00

**Whitham's** Steam-engine Design... .. ............. .... .... ....8vo, 5 00

**Wood's** Thermodynamics, Heat Motors, and Refrigerating Machines. . .8vo, 4 00

## MECHANICS PURE AND APPLIED.

**Church's** Mechanics of Engineering ...............................8vo, 6 00

  Mechanics of Fluids (Being Part IV of Mechanics of Engineering)..8vo, 3 00

  * Mechanics of Internal Work . . . .. .. 8vo, 1 50

  Mechanics of Solids (Being Parts I, II, III of Mechanics of Engineering).

8vo, 4 50

  Notes and Examples in Mechanics. .. ... .. ................8vo, 2 00

**Dana's** Text-book of Elementary Mechanics for Colleges and Schools .12mo, 1 50

**Du Bois's** Elementary Principles of Mechanics·

  Vol. I. Kinematics .... ......... ......... ........8vo, 3 50

  Vol II Statics ...... . ......... ............. .. ....8vo, 4 00

  Mechanics of Engineering. Vol I. ............. .. ...Small 4to, 7 50

  Vol. II .. . ........ .. .Small 4to, 10 00

* **Greene's** Structural Mechanics .. ... . .... ............. 8vo, 2 50

* **Hartmann's** Elementary Mechanics for Engineering Students ...12mo, 1 25

**James's** Kinematics of a Point and the Rational Mechanics of a Particle.

Large 12mo, 2 00

* **Johnson's** (W W ) Theoretical Mechanics .. ............. .... 12mo, 3 00

* **King's** Elements of the Mechanics of Materials and of Power of Transmission . ............... . .. ..... .. . ...8vo, 2 50

**Lanza's** Applied Mechanics...................... . ....... 8vo, 7 50

* **Martin's** Text Book on Mechanics, Vol I, Statics . . ... . ..12mo, 1 25

  * Vol. II. Kinematics and Kinetics... . . 12mo, 1 50

  * Vol. III. Mechanics of Materials .. . 12mo, 1 50

**Maurer's** Technical Mechanics . ............... . .. ... 8vo, 4 00

* **Merriman's** Elements of Mechanics .. .... . ..... ....12mo, 1 00

  Mechanics of Materials . .. ......... ... ....8vo, 5 00

* **Michie's** Elements of Analytical Mechanics.......... . .. 8vo, 4 00

**Robinson's** Principles of Mechanism............. .... .. 8vo, 3 00

**Sanborn's** Mechanics Problems ... ...... . Large 12mo, 1 50

**Schwamb** and **Merrill's** Elements of Mechanism. . .... .... . ... 8vo, 3 00

**Wood's** Elements of Analytical Mechanics .... . .... . 8vo, 3 00

  Principles of Elementary Mechanics. .. ......... ....... . 12mo, 1 25

## MEDICAL.

* **Abderhalden's** Physiological Chemistry in Thirty Lectures. (Hall and Defren ) . . . ........ . 8vo, 5 00

**von Behring's** Suppression of Tuberculosis. (Bolduan ).. . .... 12mo, 1 00

* **Bolduan's** Immune Sera .... ..... .. . . . ...12mo, 1 50

**Bordet's** Studies in Immunity. (Gay ) ............. 8vo, 6 00

* **Chapin's** The Sources and Modes of Infection .. Large 12mo, 3 00

**Davenport's** Statistical Methods with Special Reference to Biological Variations . . ... . . ..... ......... . ...16mo, mor 1 50

**Ehrlich's** Collected Studies on Immunity. (Bolduan.)... ........8vo, 6 00

* **Fischer's** Nephritis . :. . . ....... Large 12mo, 2 50

  * Oedema. . . .. . ..... ....... 8vo, 2 00

  * Physiology of Alimentation ............ .... .Large 12mo, 2 00

* **de Fursac's** Manual of Psychiatry. (Rosanoff and Collins.)...Large 12mo, 2 50

* **Hammarsten's** Text-book on Physiological Chemistry. (Mandel ) .. 8vo, 4 00

**Jackson's** Directions for Laboratory Work in Physiological Chemistry..8vo, 1 25

**Lassar-Cohn's** Praxis of Urinary Analysis (Lorenz ) .............12mo, 1 00

**Mandel's** Hand-book for the Bio-Chemical Laboratory.............12mo, 1 50

* **Nelson's** Analysis of Drugs and Medicines... ........ . ... 12mo, 3 00

* **Pauli's** Physical Chemistry in the Service of Medicine. (Fischer ). 12mo, 1 25

* **Pozzi-Escot's** Toxins and Venoms and their Antibodies (Cohn ). . 12mo, 1 00

**Rostoski's** Serum Diagnosis (Bolduan )... ............. ...12mo, 1 00

**Ruddiman's** Incompatibilities in Prescriptions............... ....... . 8vo, 2 00

  Whys in Pharmacy .. ...... .. . ....... ....12mo, 1 00

**Salkowski's** Physiological and Pathological Chemistry.· (Orndorff ) ....8vo, 2 50

17

Johannsen's Determination of Rock-forming Minerals in Thin Sections 8vo,
  With Thumb Index $5 00
* Martin's Laboratory Guide to Qualitative Analysis with the Blow-
  pipe. . . . . . . . . . . . . . . . . . . . . . . . . . . . . . . . . 12mo,  0 60
Merrill's Non-metallic Minerals. Their Occurrence and Uses. . . . . . . .8vo,  4 00
  Stones for Building and Decoration . . . . . . . . . . . . . . . . . .8vo,  5 00
* Penfield's Notes on Determinative Mineralogy and Record of Mineral Tests.
  8vo, paper,  0 50
  Tables of Minerals, Including the Use of Minerals and Statistics of
    Domestic Production. . . . . . . . . . . . . . . . . . . . . . . . 8vo,  1 00
* Pirsson's Rocks and Rock Minerals. . . . . . . . . . . . . . . . . . . . . .12mo,  2 50
* Richards's Synopsis of Mineral Characters . . . . . . . . . . . . . . .12mo, mor.  1 25
* Ries's Clays. Their Occurrence, Properties and Uses . . . . . . . . . . . . 8vo,  5 00
* Ries and Leighton's History of the Clay-working industry of the United
    States . . . . . . . . . . . . . . . . . . . . . . . . . . . . . . . . . . . 8vo,  2 50
* Rowe's Practical Mineralogy Simplified. . . . . . . . . . . . . . . 12mo,  1 25
* Tillman's Text-book of Important Minerals and Rocks . . . . . . . . . . . .8vo,  2 00
Washington's Manual of the Chemical Analysis of Rocks. . . . . . . . . . . .8vo,  2 00

## MINING.

* Beard's Mine Gases and Explosions. . . . . . . . . . . . . . . . . . . . . Large 12mo,  3 00
* Crane's Gold and Silver . . . . . . . . . . . . . . . . . . . . . . . . . . . . . . . . 8vo,  5 00
    * Index of Mining Engineering Literature. . . . . . . . . . . . . . . . . . .8vo,  4 00
    * 8vo, mor.  5 00
    * Ore Mining Methods . . . . . . . . . . . . . . . . . . . . . . . . . . . . . . 8vo,  3 00
* Dana and Saunders's Rock Drilling. . . . . . . . . . . . . . . . . . . . . .8vo,  4 00
Douglas's Untechnical Addresses on Technical Subjects. . . . . . . . . .12mo,  1 00
Eissler's Modern High Explosives . . . . . . . . . . . . . . . . . . . . . . . . .8vo,  4 00
Goesel's Minerals and Metals: A Reference Book. . . . . . . . . . . . 16mo, mor.  3 00
Ihlseng's Manual of Mining. . . . . . . . . . . . . . . . . . . . . . . . . . . . . .8vo,  5 00
* Iles's Lead Smelting . . . . . . . . . . . . . . . . . . . . . . . . . . . . . . . . . .12mo,  2 50
* Peele's Compressed Air Plant . . . . . . . . . . . . . . . . . . . . . . . 8vo,  3 50
Riemer's Shaft Sinking Under Difficult Conditions    (Corning and Peele )8vo,  3 00
* Weaver's Military Explosives. . . . . . . . . . . . . . . . . . . . . . . . . . . 8vo,  3 00
Wilson's Hydraulic and Placer Mining    2d edition, rewritten . . . . . .12mo,  2 50
    Treatise on Practical and Theoretical Mine Ventilation . . . . . . . .12mo,  1 25

## SANITARY SCIENCE.

Association of State and National Food and Dairy Departments, Hartford
    Meeting, 1906. . . . . . . . . . . . . . . . . . . . . . . . . . . . . . . . . . . . 8vo,  3 00
    Jamestown Meeting, 1907. . . . . . . . . . . . . . . . . . . . . . . . . . . . .8vo,  3 00
* Bashore's Outlines of Practical Sanitation. . . . . . . . . . . . . . . . 12mo,  1 25
    Sanitation of a Country House . . . . . . . . . . . . . . . . . . . . . .12mo,  1 00
    Sanitation of Recreation Camps and Parks. . . . . . . . . . . . . . . . .12mo,  1 00
* Chapin's The Sources and Modes of Infection    . . . . . . . . . .Large 12mo,  3 00
Folwell's Sewerage    (Designing, Construction, and Maintenance ) . . .8vo,  3 00
    Water-supply Engineering. . . . . . . . . . . . . . . . . . . . . . . . . . . 8vo,  4 00
Fowler's Sewage Works Analyses. . . . . . . . . . . . . . . . . . . . . . . . . . .12mo,  2 00
Fuertes's Water-filtration Works. . . . . . . . . . . . . . . . . . . . . . . . . 12mo,  2 50
    Water and Public Health. . . . . . . . . . . . . . . . . . . . . . . . . . . .12mo,  1 50
Gerhard's Guide to Sanitary Inspections. . . . . . . . . . . . . . . . . . . .12mo,  1 50
    * Modern Baths and Bath Houses. . . . . . . . . . . . . . . . . . . . . . 8vo,  3 00
    Sanitation of Public Buildings . . . . . . . . . . . . . . . . . . . . . . . 12mo,  1 50
    * The Water Supply, Sewerage, and Plumbing of Modern City Buildings.
    8vo,  4 00
Hazen's Clean Water and How to Get It . . . . . . . . . . . . . . . . . Large 12mo,  1 50
    Filtration of Public Water-supplies. . . . . . . . . . . . . . . . . . . .8vo,  3 00
* Kinnicutt, Winslow and Pratt's Sewage Disposal    . . . . . . . . . . 8vo,  3 00
Leach's Inspection and Analysis of Food with Special Reference to State
    Control . . . . . . . . . . . . . . . . . . . . . . . . . . . . . . . . . . . 8vo,  7 50
Mason's Examination of Water.    (Chemical and Bacteriological). . . . .12mo,  1 25
    Water-supply.    (Considered principally from a Sanitary Standpoint).
    8vo,  4 00
* Mast's Light and the Behavior of Organisms. . . . . . . . . . . . . . .Large 12mo,  2 50

18

Merriman's Elements of Sanitary Engineering . . .   . . . . . . . . . . . .  .. 8vo, $2 00
Ogden's Sewer Construction . . . . . . . .  . .  . . .  . . . . . . .. 8vo,  3 00
    Sewer Design . .  . .  . .  . . . . . .  . . . . . . . . . .12mo,  2 00
Parsons's Disposal of Municipal Refuse  .  . . . . . 8vo,  2 00
Prescott and Winslow's Elements of Water Bacteriology, with Special Refer-
    · ence to Sanitary Water Analysis  . . . . . .  . . . . . . 12mo,  1 50
* Price's Handbook on Sanitation  . . . . . . . . . . . . .12mo,  1 50
Richards's Conservation by Sanitation . . . . .  .  . . . . . . .8vo,  2 50
    Cost of Cleanness  .  .  . . . . . . . .  . . . . . . . . .12mo,  1 00
    Cost of Food  A Study in Dietaries. . . . .  .  . . . . . . . . . . .12mo,  1 00
    Cost of Living as Modified by Sanitary Science . . . . . . . . .  . . . 12mo,  1 00
    Cost of Shelter . .  . . .  . . .  . .  . . . . . . . . . .  .. 12mo,  1 00
Richards and Woodman's Air, Water, and Food from a Sanitary Stand-
    point . . . . . . . . . . . . . . . . .  . . . . . . . . . . . . . .8vo,  2 00
* Richey's Plumbers', Steam-fitters', and Tinners' Edition (Building
    Mechanics' Ready Reference Series)  . .  . . . . . . . . . . . .16mo, mor.  1 50
Rideal's Disinfection and the Preservation of Food. . . . .  . . . . . . . .  8vo,  4 00
Soper's Air and Ventilation of Subways . . . . . . . . . . . . . . .  . . . . . . . .12mo,  2 50
Turneaure and Russell's Public Water-supplies. . . . . . . . . . . . .  8vo,  5 00
Venable's Garbage Crematories in America . . . . . . . . . . . . . . . . . . .8vo,  2 00
    Method and Devices for Bacterial Treatment of Sewage. . . . . . . . .8vo,  3 00
Ward and Whipple's Freshwater Biology.  (In Press )
Whipple's Microscopy of Drinking-water. . . . . . . . . . . . . . . . . . . . . .8vo,  3 50
    * Typhoid Fever. . . . . . . . . . . . . . . . . . . . . . . . . . . . . . . . .Large 12mo,  3 00
    Value of Pure Water  . . . . . . . . . . . . . . . . . . . . . . . . . .Large 12mo,  1 00
Winslow's Systematic Relationship of the Coccaceæ . . . . . . . . . .Large 12mo,  2 50

## MISCELLANEOUS.

* Burt's Railway Station Service. . . .  . . . . . . . . . . .  .  . .  . . .  . . . . .12mo,  2 00
* Chapin's How to Enamel.  . .  . . . . . . . .  . . .  . . . . . .  . . . . . . 12mo,  1 00
Emmons's Geological Guide-book of the Rocky Mountain Excursion of the
    International Congress of Geologists . . . . . . . . . .  . . . . .Large 8vo,  1 50
Ferrel's Popular Treatise on the Winds. . . . . . . . . . . . . . . . . . . . . . . .8vo,  4 00
Fitzgerald's Boston Machinist . . . . . . . . . . . . . . . . . . . . . . . . . . . .18mo,  1 00
* Fritz, Autobiography of John  . . . .  . . . . . . . . .  . . . . .  .  . . . . .8vo,  2 00
Gannett's Statistical Abstract of the World. . . . . . . . . . . . . . . . . . .24mo,  0 75
Haines's American Railway Management . . . . . . . . . . . . . . . . . . . . . .12mo,  2 50
Hanausek's The Microscopy of Technical Products  (Winton) . . . . . . .8vo,  5 00
Jacobs's Betterment Briefs  A Collection of Published Papers on Or-
    ganized Industrial Efficiency. . . . . . . . . . . . . . . . . . . . . . . . . . .8vo,  3 50
Metcalfe's Cost of Manufactures, and the Administration of Workshops. .8vo,  5 00
* Parkhurst's Applied Methods of Scientific Management. . . . . . . . . .  8vo,  2 00
Putnam's Nautical Charts. . . . . . . . . . . . . . . . . . . . . . . . . . . . . . . . .8vo,  2 00
Ricketts's History of Rensselaer Polytechnic Institute 1824–1894.
                                             Large 12mo,  3 00
* Rotch and Palmer's Charts of the Atmosphere for Aeronauts and Aviators.
                                               Oblong 4to,  2 00
Rotherham's Emphasised New Testament . . . . . . . . . . . . . . . . . . .Large 8vo,  2 00
Rust's Ex-Meridian Altitude, Azimuth and Star-finding Tables. . . . . . .8vo  5 00
Standage's Decoration of Wood, Glass, Metal, etc . . . . . . . . . . . . . . . . .12mo  2 00
Thome's Structural and Physiological Botany.  (Bennett). . . . . . . . . .16mo,  2 25
Westermaier's Compendium of General Botany.  (Schneider). . . . . . . . .8vo,  2 00
Winslow's Elements of Applied Microscopy. . . . . . . . . . . . . . . . . . . . .12mo,  1 50

## HEBREW AND CHALDEE TEXT-BOOKS.

Gesenius's Hebrew and Chaldee Lexicon to the Old Testament Scriptures.
    (Tregelles ).  . . . . . .  . . . . . . . . . . . . . . . . . . . . . .Small 4to, half mor,  5 00
Green's Elementary Hebrew Grammar. . . . . . . . . . . . . . . . . . . . . . . . .12mo  1 25

www.ingramcontent.com/pod-product-compliance
Lightning Source LLC
Chambersburg PA
CBHW031920190326
41519CB00007B/355